Target Cost Contracting Strategy in Construction

T0179419

The problems inherent in the traditional design-bid-build procurement method often lead to adversarial working relationships within the construction industry. Target cost contracts, accompanied by a gain-share/pain-share arrangement serving as a cost incentive mechanism, have emerged in the United States, the United Kingdom, Australia and Hong Kong with the aim of achieving better value for money and more satisfactory overall project performance under a collaborative working relationship.

This book presents the underlying principles and practicalities of applying the target cost contracting strategy, along with a series of short case studies. Principles begin with the fundamentals, then cover the development of target cost contracting in major countries/cities, definitions of target cost contracting, perceived benefits, potential difficulties and critical success factors for implementation.

Practices include the target cost contracting approach and process in general, and the key risk factors, risk assessment model, risk allocation and risk mitigation measures for target cost contracts in particular, together with a conceptual framework for the performance measurement of target cost contracts. Several short real-life case studies from the United Kingdom, Hong Kong, Australia and New Zealand are provided for further illustration.

The book will appeal to a wide spectrum of readers, from industrial practitioners to undergraduate students, researchers and academics interested in construction contracts and procurement methods.

Daniel W.M. Chan is Associate Professor in Construction Project Management and the Associate Head (Teaching and Learning) at the Department of Building and Real Estate at The Hong Kong Polytechnic University. He has published over 200 research papers on the broad theme of project management in leading construction management journals and international conference proceedings.

Joseph H.L. Chan is Lecturer in Housing Management at the School of Professional Education and Executive Development (SPEED) at The Hong Kong Polytechnic University. To date, he has produced over twenty publications, including journal papers, international conference papers, consultancy reports and other written outputs related to construction procurement management.

Spon Research

Spon Research publishes a stream of advanced books for built environment researchers and professionals from one of the world's leading publishers. The ISSN for the Spon Research programme is ISSN 1940-7653 and the ISSN for the Spon Research E-book programme is ISSN 1940-8005.

Published:

Relational Contracting for
Construction Excellence:
Principles, practices and
case studies
978-0-415-46669-1
A. Chan, D. Chan & J. Yeung

Soil Consolidation Analysis
978-0-415-67502-4
J.H. Yin & G. Zhu

OHS Electronic
Management Systems for
Construction
978-0-415-55371-1
I. Kamardeen

FRP-Strengthened Metallic
Structures
978-0-415-46821-3
X.L. Zhao

Leadership and Sustainability
in the Built Environment
978-1-138-77842-9
A. Opoku & V. Ahmed

The Soft Power of Construction
Contracting Organisations
978-1-138-80528-6
*S.O. Cheung, P.S.P. Wong &
T.W. Yiu*

Fall Prevention through Design
in Construction: The benefits of
mobile construction
978-1-138-77915-0
I. Kamardeen

Trust in Construction Projects
978-1-138-81416-5
A. Ceric

New Forms of Procurement:
PPP and relational contracting in
the 21st century
978-1-138-79612-6
M. Jefferies & S. Rowlinson

Target Cost Contracting Strategy
in Construction: Principles,
practices and case studies
978-1-315-62453-2
D.W.M. Chan & J.H.L. Chan

Target Cost Contracting Strategy in Construction
Principles, practices and case studies

Daniel W.M. Chan and
Joseph H.L. Chan

Sponsored by HKIS

THE HONG KONG INSTITUTE OF
SURVEYORS
香 港 測 量 師 學 會

Routledge
Taylor & Francis Group

LONDON AND NEW YORK

First published 2017 by Routledge

2 Park Square, Milton Park, Abingdon, Oxon, OX14 4RN
605 Third Avenue, New York, NY 10017

Routledge is an imprint of the Taylor & Francis Group, an informa business

First issued in paperback 2020

British Library Cataloguing in Publication Data
A catalogue record for this book is available from the British Library

Library of Congress Cataloging-in-Publication Data
Names: Chan, Daniel W. M. | Chan, Joseph H. L.
Title: Target cost contracting strategy in construction : principles, practices and case studies / Daniel W.M. Chan and Joseph H.L. Chan.
Description: Abingdon, Oxon ; New York, NY : Routledge, 2017. | Series: Spon research, ISSN 1940-7653 | Includes bibliographical references and index.
Identifiers: LCCN 2016022086| ISBN 9781138651906 (hardback : alk. paper) | ISBN 9781315624532 (eBook)
Subjects: LCSH: Construction industry—Cost control. | Target costing. | Construction contracts.
Classification: LCC TH438.15 .C43 2017 | DDC 692/.8—dc23
LC record available at https://lccn.loc.gov/2016022086

ISBN: 978-1-138-65190-6 (hbk)
ISBN: 978-0-367-73674-3 (pbk)

Typeset in Sabon
by Swales & Willis Ltd, Exeter, Devon, UK

Contents

PART III
Case studies in target cost contracts **119**

Figures

Tables

Foreword

The construction industry is regarded as a very competitive and risky business. The fixed-price lump-sum contractual arrangement is a major contributor to such perception. Weaknesses of this contractual arrangement include limited mutual trust among project stakeholders and a lack of incentives and common goals. Confrontational working relationships could arise easily which would result in unfavourable project performance.

Target Cost Contracting (TCC) has emerged in recent years as a sound contracting model. It can bring significant benefits to all contracting parties in terms of both effectiveness and efficiency, as it builds upon a co-operative, trust-based relationship and partnering spirit with a gain-share/pain-share mechanism in place. The authors share with readers their personal hands-on experiences and useful findings from previous projects researching TCC over the past decade.

The book covers the principles and practices of TCC and the case studies give readers a good reflection of the real world. Regarding the principles of TCC, the authors examine the fundamental concepts, historical development in major countries/cities, perceived benefits, potential difficulties and critical success factors. The well-conceived discussions of risk management and performance measurement reveal how principles are put into practice. An in-depth account of risk management has been provided, including the identification of key risk factors, the risk assessment model, risk allocation and risk mitigation. Performance measurement is explained by using a conceptual framework for identification of key performance indicators (KPIs). The real-life case studies in London, Hong Kong, South Australia and New Zealand are invaluable and informative reference materials for readers.

I recommend this book to undergraduate and postgraduate students, university academics and construction professionals who would like to learn more about the theory and application of Target Cost Contracting.

Sr Lau Chun-Kong
President (2015–2016)
The Hong Kong Institute of Surveyors
July 2016

Preface

The problems inherent in the traditional design-bid-build procurement method often lead to the adversarial working relationships within the construction industry. Target cost contracts, accompanied by a gain-share/pain-share arrangement serving as a cost incentive mechanism, have emerged in the United States, the United Kingdom, Australia and Hong Kong with the aim of achieving better value for money and more satisfactory overall project performance under a collaborative working relationship. Moreover, target cost contracts are considered to be a desirable procurement option where project risks are taken jointly by the employer and the contractor.

Based on the authors' ten years of research and teaching experience in target cost contracts, this book presents the underlying principles, practicalities and a series of short case studies of applying the target cost contracting strategy. Principles begin with the fundamentals, then cover the development of target cost contracting in major countries/cities, definitions of target cost contracting, perceived benefits, potential difficulties and critical success factors for implementation.

Practices include the target cost contracting approach and process in general, and significant factors which make target cost contracts successful, the key risk factors, risk assessment model, risk allocation and risk mitigation measures for target cost contracts in particular. A conceptual framework for the performance measurement of target cost contracts is also indicated in this book. A plethora of short real-life case studies from the United Kingdom, Hong Kong, Australia and New Zealand are provided for further illustration.

Since the book draws on a combination of practical case studies from the industry and research outcomes from academia, a wide spectrum of readers will find it useful and informative, from industrial practitioners to undergraduate students.

The authors would like to acknowledge the concerted efforts made by the relevant research team members, including Professor Albert P.C. Chan and Dr Patrick T.I. Lam (The Hong Kong Polytechnic University), together with Mr Wayne Lord (Loughborough University) and Dr Tony Ma (University of South Australia), for their unfailing support and generous participation.

Sincere thanks are also extended to the part-time student assistant, Miss Hannah Yeung, for offering technical support in drafting this book for publication.

Moreover, the authors wish to express their heartfelt gratitude to the Board of Education of the Hong Kong Institute of Surveyors for providing financial sponsorship to the compilation of this book. Lastly, they are very grateful for the financial support from the Research Grants Council (RGC) of the Hong Kong Special Administrative Region, the Faculty of Construction and Environment and the Department of Building and Real Estate of The Hong Kong Polytechnic University (HK PolyU), in funding the following research projects upon which the contents of this book are based:

- An Investigation of Guaranteed Maximum Price (GMP) and Target Cost Contracting (TCC) Procurement Strategies in Hong Kong Construction Industry (HK PolyU Faculty Internal Competitive Research Grants Allocation 2004–2005)
- Exploring the Key Risk Factors and Risk Sharing Mechanisms for Guaranteed Maximum Price (GMP) and Target Cost Contracting (TCC) Schemes in Hong Kong (HK PolyU Departmental General Research Grants Allocation 2006–2007)
- Evaluating the Key Risk Factors and Risk Sharing Mechanisms for Target Cost Contracting (TCC) Schemes in Construction (RGC General Research Fund 2007–2008)
- Development of an Overall Performance Index (PI) for Target Cost Contracts in Construction (RGC General Research Fund 2008–2009 for Proposal Rated 3.5)

Dr Daniel Chan and Dr Joseph Chan
The Hong Kong Polytechnic University
July 2016

Abbreviations

DLS	Davis Langdon and Seah
ECC	Engineering and Construction Contract
FIDIC	Fédération Internationale Des Ingénieurs-Conseils
FRAM	fuzzy risk assessment model
GMP	guaranteed maximum price
HK PolyU	The Hong Kong Polytechnic University
HKSAR	Hong Kong Special Administrative Region
IA	incentivisation agreement
KCRC	Kowloon–Canton Railway Corporation
KPI	key performance indicator
MTRC	Mass Transit Railway Corporation
NEC	New Engineering Contract
NHS	National Health Service (United Kingdom)
ODA	Olympic Delivery Authority
ORI	Overall Risk Index
PMI	performance measurement index
PRF	principal risk factor
PRG	principal risk group
RGC	Research Grants Council of the Hong Kong Special Administrative Region
TCC	target cost contracting/target cost contract

Part I
Principles of target cost contracting

1 Fundamentals of target cost contracts

Introduction

The selection of procurement methods for construction projects is a critical factor in achieving project success (Chan and Yung, 2003). The construction industry keeps changing, resulting in more belief that it is essential that procurement methods are tailor-made in order to align the goals and aspirations of various key project stakeholders, agree on the risk allocation mechanism, and decide on what portion of the design and construction the client intends to be integrated. Chapter 1 sets out the basics of TCC and GMP in construction projects.

Definitions and characteristics of TCC

Target cost contracting (TCC) and guaranteed maximum price (GMP) are an incentive-based strategy for construction procurement. According to Masterman (2002), TCC and GMP award contractors for the cost savings against the target cost or guaranteed price, but penalise them when the cost exceeds the agreed amount due to their own faults in terms of management or negligence pursuant to the pre-agreed share ratio. The contractor usually includes a TCC and GMP allowance to cover future design development and unexpected risks (Gander and Hemsley, 1997).

Target cost contracting

TCC can be defined as follows:

> Target cost contracts specify a best estimate of the cost of the work to be carried out. During the course of the work, the initial target cost will be adjusted by agreement between the client or his nominated representative and the contractor to allow for any changes to the original specifications. Any savings or overruns between target cost and actual cost at completion are shared between the parties to the contract.
>
> (National Economic Development Office, 1982)

Trench (1991) also stated that under TCC, after evaluating and comparing the actual cost of the completion of construction with the target cost of the project, the client and contractor share the differences within a cost band. The Mass Transit Railway Corporation (MTRC, 2003) also commented that 'the client and the contractor would share savings (gains) if the final account figure turns out to be less than the target. Should the final account exceed the target, they would share the excess (pain).' This is different from a fixed-price contract due to the agreement and collective determination of the sharing ratio of risks.

Wong (2006) also explained that the builder will be paid the actual cost for the work completed throughout construction. When there is a difference between the initial contract target cost and the actual cost, the client and contractor will share the difference based on the pre-agreed gain-share/pain-share proportion. Hughes et al. (2011) also stated that TCC is often regarded as a gain-share/pain-share approach in which the contracting parties determine a target cost (estimated cost) and sharing ratio which is used when the estimated cost is higher or lower than the actual cost. They also suggested that TCC is suitable if: (1) the client is incentivised to become actively involved to help the contractor to develop cost-efficient solutions; and (2) the same contractor is deliberately chosen for recurrent business by the client.

Guaranteed maximum pricing

According to Boukendour and Bah (2001), GMP is regarded as a hybrid procurement method involving a call option for a fixed-price contract as well as a cost imbursement contract. The contractor guarantees that the project will be completed within the pre-agreed GMP at main contract award and within the contract period in consonance with the specifications and drawings.

Carty (1995) described GMP as follows:

> The contractor and owner agree that the contractor will perform an agreed scope of work (defined as best as possible) at a price not to exceed an agreed upon amount, the guaranteed maximum price (GMP) . . . if these costs and the agreed upon contractor's profit are less than the GMP, the owner and contractor will share the savings in cost based upon an agreed upon formula. If the costs exceed the GMP without any changes to the defined scope, the contractor must solely bear the additional cost.

Kerzner (1995) elaborated on Carty (1995)'s definition of GMP as follows:

> the contractor is paid a fixed fee for his profit and reimbursed for the actual cost of engineering, materials, construction labour, but only up to the ceiling figure established as the 'maximum guaranteed'.

Savings below the maximum guaranteed are shared between owner and contractor, whereas the contractor assumes the responsibility for any over-run beyond the guaranteed maximum price.

(Kerzner, 1995, cited in Ferreira and Rogerson, 1999)

As mentioned by Perry and Thompson (1982), GMP could thus be regarded as a form of TCC where the shared proportion of GMP is restricted only to the gain.

Procurement route of TCC

Tendering method

If a project is procured under TCC and GMP on a negotiated contract basis, a suitable builder will already have been selected by means of a corporate

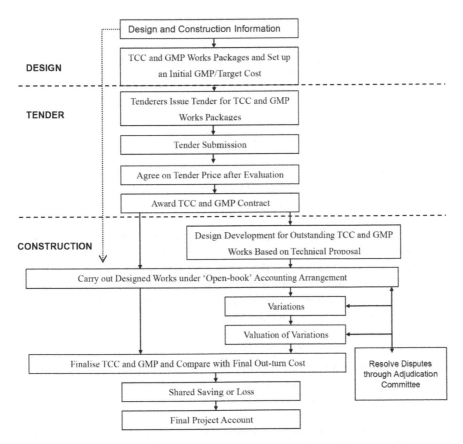

Figure 1.1 TCC and GMP contract procurement route (Hong Kong Housing Authority, 2006).

relationship between the companies. Especially in Hong Kong, most of the GMP contracts are awarded to a favourable contractor on a negotiated basis owing to a corporate business relationship – e.g. Hong Kong Land Ltd (developer) working with Gammon Construction Ltd (main contractor). On the other hand, in selective tendering, the client will invite the tenderers to pre-qualify by submitting a preliminary proposal including details of company strength, previous track record, related project experience, financial soundness, capability in alternative procurement methods, technical competence, partnering dedication, organisational structures and staff. Eventually, the employer and the employer's team of consultants review the proposals. After a thorough evaluation, the client will shortlist and invite the pre-qualified contractors to submit a tender. Figure 1.1 demonstrates a typical procurement route under the TCC and GMP approaches.

If a two-stage tendering method is adopted, after pre-qualification, tenderers will be invited to submit tenders in accordance with the following preliminary materials provided by client and the client's team of consultants:

1   cost plan;
2   base schematic/outline design drawings (e.g. ~20% of design complete);
3   performance specifications for works packages;
4   other accessible information.

After evaluating the tenders, the tenderers will be shortlisted, then requested to submit more detailed proposals in the second stage, consisting of:

1   a bill of quantities;
2   a more detailed set of design drawings;
3   performance specifications for works packages.

A negotiated tendering arrangement will not diminish the purposes of gaining a competitive tender, since an 'open-book' competitive tender approach is ultimately applied in most of the subcontract packages. Nevertheless, this information exchange demands a high level of mutual trust among the contracting stakeholders, particularly the main contractor. The amount of the competitive subcontract packages tendered may be 60–80% of the contract value.

Regarding the materials required for the TCC and GMP contracts, the employer and team of consultants provide the preliminary design documentation for estimating both the target cost and GMP. Tender documents for GMP contracts usually consist of the following:

1   cost for main contractor's direct works (e.g. substructure works, reinforced concrete superstructure works, finishes works, etc.);
2   domestic subcontractor's works packages;

3   provisional quantities;
4   provisional sums;
5   design development allowance (Hong Kong Housing Authority, 2006).

Figure 1.2 lists the details of these documents. As the information stated in the tender documents may be insufficient to construct and complete the project, the builder is allowed to include a sum for design development in the tender price. After the target cost is agreed and conveyed to the contractor with the architect's instructions, the employer and consultant team will provide more detailed design information to the contractor.

Generally, the tender documents for domestic subcontractors' works packages (e.g. plumbing and drainage, electrical and mechanical installation, air conditioning installation, fire services installation, lift installation, specialist external works, etc.) are prepared by the main contractor with the team of consultants. The preferred or pre-qualified subcontractors will receive tender documents that dictate the quality and state the range of work. The main contractor needs to show any GMP variations that will require recalculating the GMP in the subcontract tender documents. After issuing the subcontract tender documents to the suitable tenderers, the scope of work stated in the tender document for the subcontractor's works package

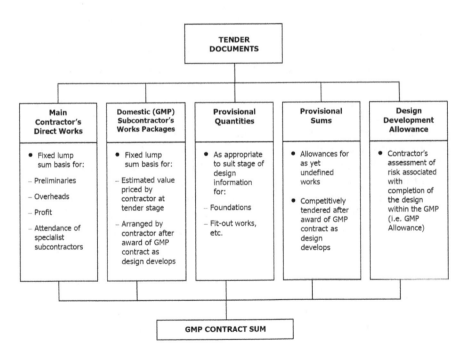

Figure 1.2 Tender documents for a typical GMP project (Hong Kong Housing Authority, 2006).

will be deemed to have been accepted by the main contractor and involved in the allowances for future design development, which means that GMP will not be recalculated. The main contractor and the employer's team of consultants will evaluate the tenders and give further advice to the employer for contract award on a competitive 'open-book' strategy, which should ensure that the subcontractors are evaluated fairly on the tender sum they provided.

The successful subcontractor will enter into a domestic subcontract with the main contractor. Such a procedure will eliminate the need for nominated subcontracts and the underlying liabilities. In addition, the main contractor can be assured that the subcontractor can only subcontract or assign works with the client's approval. The savings, if any, acquired in the tendering of subcontractors are included in the final out-turn costs and become the basis for calculating the shared saving at completion stage.

Pricing mechanism

Similar to other standard cost-based procurement contracts, the main contractor's tender pricing of subcontract packages should be available to the client in detail. This may be achieved through an 'open-book' accounting method. The employer may scrutinise the contractor's account to ensure that a strong administrative team and the contractor's accountant are involved on site. If the client is satisfied with the built facilities, it will pay these costs to the main contractor. According to the National Economic Development Office (1982), improved responsibility and quantification of risks can be achieved through using open-book accounting schemes.

Another feature of TCC and GMP contracts is that the completion will be within the target cost, as warranted or promised by the main contractor (Gander and Hemsley, 1997). Under GMP procurement, if the actual cost exceeds the negotiated GMP, the client may merely pay the GMP amount and the contractor will be liable for the excess costs (Cantirino and Fodor, 1999). Therefore, the client can moderate the financial risk by establishing this price ceiling (Boukendour and Bah, 2001).

The unique character of TCC is the gain-share/pain-share mechanism (Trench, 1991). If the actual cost and the target cost are different, TCC enables the main contractor and the client to share the 'gain/pain'. TCC thereby generates strong incentives for contractors to minimise costs by applying their own experience, expertise and innovations at both the design and construction stages. Figure 1.3 shows an example under this gain-share/pain-share regime for both TCC and GMP projects.

The open-book accounting approach conceives that the main contractor is required to furnish all the data and information involving:

- construction programmes, method statements and resource programmes;
- pricing of the preliminaries and contract conditions;

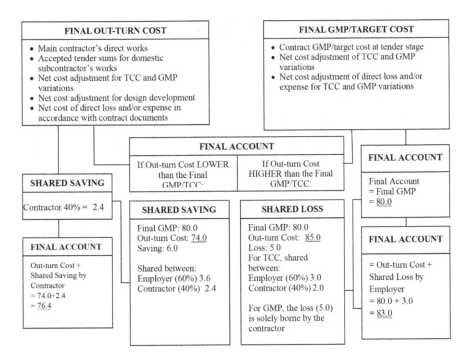

Figure 1.3 Example to illustrate the gain-share/pain-share philosophy of the TCC and GMP approach (Hong Kong Housing Authority, 2006).

- details of pricing obtained for the domestic subcontractor's works packages;
- details of attendance for subcontractors;
- details of the main contractor's direct works;
- a detailed breakdown and calculations for overheads and profit.

This information is provided to the project team members on an open-book basis in supporting the tender price during negotiation of TCC and GMP.

Contractor's inputs in design and construction

TCC and GMP are considered to be a combination of conventional design-bid-build and design-and-build contracts by Fan and Greenwood (2004). Figure 1.4 portrays the differences among these three procurement methods. Innovation and expertise in design and construction can be brought in under TCC and GMP contracts (Masterman, 2002). Although TCC and GMP and design-and-build contracts may utilise the contractor's expertise, clients under TCC and GMP can also be involved more in the design and project cost stages and integrate the contractor's technical competence and innovations within a pre-agreed framework.

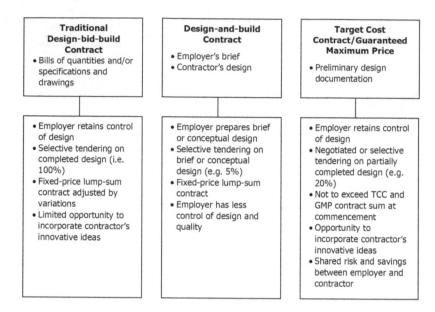

Figure 1.4 Comparisons of three alternative procurement methods (Hong Kong Housing Authority, 2006).

Project variations (consultant's instructions)

Gander and Hemsley (1997) divided variations in TCC and GMP projects into two main types: (1) design development variations (i.e. non-TCC and GMP variations), and (2) TCC and GMP variations.

Regarding design development variations, no re-calculation of target cost or GMP can be triggered, as these variations are considered to be included in the lump sum of direct works completed during the tender stage. On the contrary, re-calculation of GMP and target cost is allowed for TCC and GMP variations (Fan and Greenwood, 2004; Hong Kong Housing Authority, 2006), which could only arise under the following conditions:

1   changes in scope of work, e.g. increase/decrease in floor area or volume;
2   change in function of an area;
3   change in quality of an area;
4   adjustment of provisional quantities or provisional sums;
5   corrected quantity errors by consultants;
6   unforeseen additional fees or charges imposed by statutory authorities (Fan and Greenwood, 2004).

Therefore, extras have to be requested by the client relating to the scope changes. These will be evaluated according to the schedule of rates stated in the contract document and measured works. The net cost adjustment of such

TCC and GMP variations will be subtracted from (for 'omission' work) or added to (for 'addition' work) the target cost or GMP stated in the contract.

If the contractor is not content with the architect's decision on whether or not to classify the claim as a TCC and GMP variation, or wants to lodge a claim beyond a TCC and GMP variation, it will notify the architect in writing and offer advice on the value and extension of time, if any. All these should be conducted based on the agreed TCC and GMP methodology. In addition, if the contractor and the architect cannot agree on the definition of a TCC and GMP variation, a meeting of an adjudication committee, comprising representatives from the employer, architect, quantity surveyor and main contractor, should be convened by the architect in order to further evaluate the nature and extent of the variation as well as to help resolve any remaining problems (Hong Kong Housing Authority, 2006). This aims to identify any issues at source in order to increase efficiency and accountability. All contracting parties sharing a strong commitment and willingness is likely to lead to the success of TCC and GMP, and the prospects of this are improved by a spirit of teamwork and co-operation among all project team stakeholders (Tay et al., 2000), for instance via partnering.

Methods of setting a target cost value

There have been a plethora of research studies on setting a target cost value. For instance, Heaphy (2011) and EC Harris (2011) suggested that a TCC should consist of:

- the base cost;
- the contractor's fee;
- the risks the builder bears in the project.

According to EC Harris (2011), the base cost would involve:

- the measured work, i.e. works done by the builder, based on historical unit rates or through a resource loaded system; the builder can choose to break this down into its components (e.g. labour, plant and materials);
- the cost of all temporary works to be completed by the builder;
- any components of works to be provided by subcontractors, and fixed and time-related preliminary costs.

If the base cost is priced net of risks or allocated risks, then this has to be clearly identified, as this will ensure that there will be no double counting when subsequent allowances for the builder's risks are established (EC Harris, 2011).

The contractor's fee is the sum of:

- head and regional office overheads;
- head and regional office staff;
- insurance;
- profit (EC Harris, 2011).

However, it cannot be calculated easily or accurately on an actual cost basis (EC Harris, 2011).

Regarding risks, there should be an allowance for them. Builders should price for the cost and time effects of risks if they occur. The client's risks are related to the opportunities for the builder to claim for an extension of time and/or additional payment (e.g. variations) (EC Harris, 2011).

Another approach suggested by Zimina et al. (2012) is that the target cost for TCC projects is a sum of the variables (i.e. allowable, anticipated, market cost, other possible quantitative relationships). It is important to arrive at the 'target value design' through co-operation to minimise the construction cost.

Cost adjustment mechanism of TCC

Figure 1.5 illustrates the definitions and the cost adjustment mechanisms of TCC and GMP. Under the agreement, a ceiling price and a gain-share/pain-share

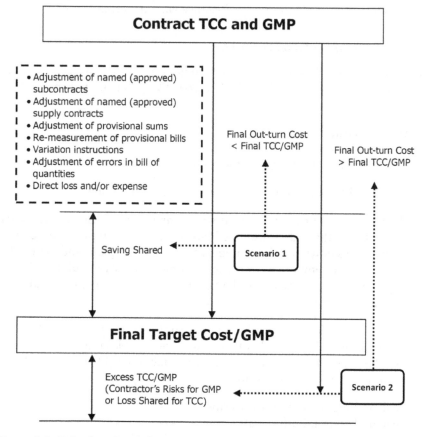

Figure 1.5 Gain-share/pain-share mechanism of TCC and GMP procurement strategies (Chan et al., 2007a) (with permission from The Hong Kong Polytechnic University).

mechanism are established in the construction contract (Clough and Sears, 1994; Patterson, 1999; Cantirino and Fodor, 1999).

Chapter summary

The following salient features of TCC and GMP procurement strategies are summarised by Chan et al. (2006):

- An agreed ceiling price is set at the main contract award for the client under the GMP procurement strategy.
- The project is guaranteed to be completed within the period agreed in the contract since the contractor may start construction early, before the design is completely developed.
- The contractor may utilise its expertise to provide advice on building designs, construction methods and innovations during the tender and post-tender stages in order to increase the buildability of the construction project by submitting alternative proposals.
- The gain-share/pain-share approach introduces financial incentives for the contractor to maximise cost savings in the pre-agreed sharing ratio after the main contract award by driving the procurement process efficiently.
- The client may transfer all the risks which are likely to be incurred during design development to the contractor by stating a TCC and GMP allowance in the tender.
- Since project design will not be thoroughly developed during tendering, the contractor may price for design development.
- An adjudication committee is established to facilitate the resolution of contentious issues, including representatives from the employer, architect, main contractor and quantity surveyor.
- The client has greater control over and influence on the design team, main contractor and subcontractors.
- Mutual goals are established to develop an integrated and trustful working team with a partnering approach. Since the time and price implications of any potential changes are pre-agreed, the final account may be settled early.
- As the details of the contractor's pricing are based on an open-book accounting approach, this may improve the transparency of the project cost, variations and quantification of the costs of risks.

2 Target cost contracting

Development in major countries/cities

Introduction

Lack of mutual trust between clients and contractors, inadequate co-operation and misalignments of common goals are the main causes of a confrontational working relationship among project stakeholders, impairing the overall project performance (Kaka et al., 2008). Many industry review reports (e.g. Construction Industry Review Committee, 2001; Egan, 1998; Latham, 1994) have stated that clients in both the private and public sectors were dissatisfied with the traditional procurement method.

Alternative integrated procurement strategies are needed to overcome the problems of fragmented working relationships and insufficient incentives for project stakeholders to contribute more than the minimum contractual obligations under the conventional design-bid-build arrangement (Latham, 1994; Construction Industry Review Committee, 2001). TCC and GMP schemes for aligning the individual goals have been adopted in the United Kingdom, Australia, Hong Kong and the United States in order to achieve more desirable project outcomes (Trench, 1991; Chan et al., 2007b; Bogus et al., 2010).

In this chapter, the history and the development of TCC and GMP schemes in the United Kingdom, Australia and Hong Kong are explored and studied.

History of the development of TCC in construction

According to the United Kingdom National Health Service (NHS) *Executive Summary of ProCure21+* (NHS, 2010), TCC was widely implemented in the construction of health services premises by the Department of Health of the United Kingdom. Before June 2010, 317 projects were completed under Option C (TCC with activity schedule) of the New Engineering Contract Version 2 (NEC2). The NHS (2010) reported that both time and cost were lower than with traditional procurement contracts in the United Kingdom.

In Hong Kong, some construction projects have applied TCC, especially for the infrastructure sector, over recent years. For instance, TCC was adopted in the Tsim Sha Tsui Metro Station Modification Works. This project was accomplished earlier than the date expected and within budget (Chan et al., 2010b). The civil engineering works contracts of the South Island Railway Line started in late 2010 also utilised the same procurement strategy (*Hong Kong Engineer*, 2011). It is worth noting that all these TCC-procured projects are civil engineering works projects. Regarding the private sector, property developers tend to use GMP contracts (Chan et al., 2007a), as the expenditure of developers is capped if there is no change in GMP value. Sometimes, if the GMP procurement method is adopted for a civil works project, the contractor may intend to price for a higher risk premium. The client may not be willing to accept the higher risk premium to cover, for example, the risks of unforeseen ground conditions and variations in the scope of works.

United Kingdom

The TCC and GMP procurement approaches have attracted the attention of industrial practitioners worldwide over the past ten years. In the United Kingdom, Mylius (2007) reported that the New Wembley Stadium in London, procured under the GMP arrangement, was opened in March 2007. Its cost exceeded £757 million, which is higher than the initial estimate of £200 million in 1996, and it was delayed by almost two years. Meng and Gallagher (2012) examined the project performance, involving cost, time and quality, of sixty completed construction projects in the United Kingdom and the Republic of Ireland by conducting a questionnaire survey. Meng and Gallagher (2012) concluded that the performance of fixed-price contracts is more satisfactory than TCC. In addition, 70% of surveyed projects procured under fixed-price contracts might achieve cost savings or be completed on budget, but only around half of the projects administrated with TCC were completed under or on budget.

In Europe, Nicolini et al. (2000) investigated two large-scale construction projects implementing TCC where the costs of some items were lowered because of the adoption of innovative construction methods and solutions, thus stating that TCC may help to integrate the supply chain, and increase the quality and profitability of construction projects in the United Kingdom. This study also discovered that the relationship among project stakeholders was less adversarial under TCC contracts. Bresnen and Marshall (2000) investigated six construction projects using TCC in the United Kingdom, and found that incentives can strengthen commitment and develop mutual trust between project stakeholders in the long term. Nevertheless, material changes and inconsistencies among internal policies and stakeholders can prevent any trust developed from being sustained.

Pryke and Pearson (2006) scrutinised construction projects in France and the United Kingdom concerning gain-share/pain-share arrangements with a major procurement method and the adoption of GMP in the standard

Table 2.1 Some recent research studies published related to TCC and GMP
contracts between 2000 and March 2002 (Chan et al., 2010a) (with
permission from Sweet & Maxwell)

Authors	Year	Journal	Country	Focus
Nicolini et al.	2000	*British Journal of Management*	UK	Two case studies of TCC in the United Kingdom
Perry and Barnes	2000	*Engineering, Construction and Architectural Management*	UK	Tender evaluation of TCC
Bresnen and Marshall	2000	*Construction Management and Economics*	UK	Six case studies of construction projects with TCC
Boukendour and Bah	2001	*Construction Management and Economics*	UK	Analysis of GMP with option pricing theory
Broome and Perry	2002	*International Journal of Project Management*	UK	Determination of sharing ratios of TCC with utility theory

form of building contract. The utilisation of GMP could change contractors'
attitudes when managing variations because the contractors were more proactive in controlling the financial aspects of inappropriate variations (Pryke
and Pearson, 2006).

Many research studies were published between 2000 and March 2011
in different journals and conference proceedings relating to TCC and GMP
contracts, as shown in Table 2.1.

Australia

Many construction projects in South Australia have adopted the GMP procurement method. Some current and completed projects are displayed and
analysed in Table 2.2 (Perkin, 2008). These projects can be found in many
active industry organisations' websites and publications.

Table 2.2 shows a wide range of projects where the GMP procurement
method has been adopted, including commercial, residential, industrial, recreational and retail projects. The use of GMP also extends to multi-purpose
construction (i.e. a combination of two or more of these project types). The
duration of the listed projects ranges from 40 to 144 weeks. The contract
value varies from AU$4.4 million to AU$84 million. The types of GMP
contract consist of fixed-price lump-sum, negotiated, and design-and-build.
Design-and-build is the most common of the delivery systems. Although the
GMP procurement method has been developed since 1997, the majority of
projects utilising GMP have been completed within the four-year period

Table 2.2 Published complete or current GMP projects in South Australia (Perkin, 2008)

Project title	Location	Client	Contractor	Contract value	Delivery system	Project nature	Contract time	Completion date
Central City Bus Station	Adelaide, South Australia	West Central City Bus Terminal	Hansen Yuncken	$28 million	Design & Construct GMP	Multi-purpose (residential/ commercial)	63 weeks	August 2007
Munno Para Shopping Centre	Smithfield, South Australia	Chapley Nominees Pty Ltd	Hansen Yuncken	$27 million	Negotiated GMP	Commercial	52 weeks	October 2003
SA Water Building (VS1)	Victoria Square, Adelaide	Catholic Archdiocese of Adelaide	Hansen Yuncken	$84 million	Design & Construct GMP	Commercial	144 weeks	October 2008
SA Water fit-out	Victoria Square, Adelaide	SA Water	Hansen Yuncken	$34 million	Design & Construct GMP	Commercial (fit-out)	136 weeks	October 2008
Liberty Tower Apartments	Glenelg, South Australia	Urban Construct	Baulderstone Hornibrook	$78 million	Design & Construct GMP	Residential	116 weeks	November 2004
151 Pirie Street	Adelaide	Charter Hall/Pivot	Built Environs	$23 million	Design & Construct GMP	Commercial	54 weeks	March 2006
Paradise Wirrina Cove (food and beverage upgrade)	Wirrina Cove, South Australia	MBFi Resorts Pty Ltd	Built Environs	$4.5 million	GMP	Multi-purpose (retail/ commercial)	116 weeks	February 1997
Burnside Village	Burnside, South Australia	Burnside Village Shopping Centre	Built Environs	$4.4 million	GMP	Multi-purpose (retail/ commercial)	55 weeks	May 1997

(continued)

Table 2.2 (continued)

Project title	Location	Client	Contractor	Contract value	Delivery system	Project nature	Contract time	Completion date
Hirotec Automotive Closure Plant	Hewittson Road, Elizabeth West	Hirotec Corp.	Built Environs	$5.5 million	AS4000 amended by special conditions – GMP	Industrial	Unknown	May 2005
Rendezvous Allegra Hotel	Adelaide	Maittioli Constructions	Built Environs	$20 million	AS4000 – GMP	Residential	68 weeks	June 2002
Wine Export Facility	Out Harbour, South Australia	MTAA Superannuation Fund Property Pty Ltd	Built Environs	$13 million	Negotiated GMP	Industrial	114 weeks	August 2004
Wayville Showgrounds	Goodwood Road, South Australia	Adelaide Showgrounds Group	Built Environs	$35 million	Design & Construct GMP	Recreational – special use	50 weeks	September 2008
Harvey Norman	Port Road, Charles Sturt Industrial Estate	ISPT	Built Environs	$8 million	Design & Construct GMP	Industrial/ commercial	40 weeks	November 2008
Gepps Cross – Home HQ	Gepps Cross, South Australia	Axiom Properties Ltd	Built Environs	To be announced	Design & Construct GMP	Commercial	Current	Yet to be completed

from 2004 to 2008. Therefore, the use of GMP is a strong trend in the construction industry.

It can be noted that GMP is extensively used in a variety of projects with different values, timeframes and locations, as indicated in Table 2.2. Moreover, the utilisation of GMP becomes prevalent via the preferred delivery system – design and construct.

Hong Kong

In Hong Kong, TCC and GMP are relatively novel, but there have recently been some projects under TCC in the private sector and quasi-government sector (Chan et al., 2007a), demonstrating that TCC is still at a developing stage. In respect of the public sector, the Hong Kong Special Administrative Region (HKSAR) Government introduced the first NEC3 Option C (TCC with activity schedule) on a pilot basis to an open nullah improvement works project in Sai Kung which was managed by the Drainage Services Department in 2009 (Cheung, 2008). Hong Kong infrastructure projects, such as the Tsim Sha Tsui Metro Station Modification Works, Tung Chung Cable Car Project and the Tseung Kwan O Railway Extension, have adopted TCC (Chan et al., 2010b), as have the West and South Island Railway Lines.

GMP has also been used for some private residential and office buildings in Hong Kong (Chan et al., 2007a). Nevertheless, TCC and GMP cannot guarantee that every project is equitably successful in respect of cost, time and quality. For example, Bogus et al. (2010) stated that GMP contracts are less likely to have schedule or cost overruns than lump-sum contracts in public water and wastewater facilities projects in the United States. Chan et al. (2010b) investigated an underground railway station modification works project under the TCC strategy, showing that the TCC contractual arrangement could help achieve both time and cost savings. However, on the contrary, Rojas and Kell (2008) showed that only six of the twenty-four public school projects they studied were accomplished at or less than the contract GMP value. This finding thus demonstrated that the outcomes of GMP may not be truly 'guaranteed'.

Fourteen construction projects were studied in the questionnaire survey conducted by Chan et al. (2007a). It is noted that residential and commercial projects tend to adopt the GMP procurement approach, while the infrastructure projects utilise TCC, as shown in Table 2.3. This difference might be attributed to the higher potential risk nature of infrastructure, whereas GMP might be inadequately fair to the contractor, who must take on high risks without the existence of a pain-share mechanism. However, the selection of either TCC or GMP relies on the employer's experience and preferences. Two leading private property developers (Hong Kong Land Ltd and Swire Properties Ltd), as well as the railway transportation service provider (Mass Transit Railway Corporation Ltd) are playing a leading role in introducing TCC and GMP procurement arrangements to their construction

Table 2.3 TCC and GMP practices of the surveyed projects (I) (Chan et al., 2007a) (with permission from The Hong Kong Polytechnic University)

	Start date	Completion date	Payment mechanism		Tendering method		Who introduced TCC/GMP?		At what stage was it decided to introduce TCC and GMP?				
			GMP	TCC	Selective	Negotiated	Client	Main contractor	Feasibility	Outline design	Detailed design	Complete design	Construction design
Residential													
i. The Orchards	Aug. 2001	Sept. 2003	✓		✓		✓				✓		
ii. Public Housing at Eastern Harbour Crossing Site Phase 4	Jun. 2006	Jun. 2009	✓¹		✓		✓			✓			
Commercial													
i. 1063 King's Road	Nov. 1997	Aug. 1999	✓			✓	✓			✓			
ii. Chater House	Oct. 2000	Jul. 2002	✓			✓	✓			✓			
iii. Alexandra House Refurbishments	Nov. 2002	Nov. 2003	✓		✓		✓				✓		
iv. Tradeport Hong Kong Logistics Centre	Jul. 2001	Dec. 2002	✓		✓		✓			✓			
v. Landmark Redevelopment Phase 6 (York House)	Jan. 2005	Oct. 2006	✓			✓	✓		✓				
vi. Three Pacific Place	Jun. 2002	Aug. 2004	✓		✓		✓						✓

vii.	TKO Technology Park	Nov. 2001	Dec. 2002	✓		✓	✓
viii.	Wynn Resorts, Macau	Jun. 2004	Aug. 2006	✓		✓	
ix.	Omni Berkshire Place Hotel Renovation, New York	Jan. 1996	Oct. 2004	✓		✓	
Infrastructure							
i.	MTRC Tsim Sha Tsui Railway Station Modification Works	Apr. 2002	Sep. 2005	✓	✓	✓	
ii.	MTRC Contract 604 – Yau Tong Station	Jan. 2001	Jun. 2002	✓	✓	✓	
iii.	MTRC Contract – Tung Chung Cable Car	Jun. 2004	Dec. 2005	✓	✓	✓	

Note:

1 According to an interviewee from the Hong Kong Housing Authority and relevant tender documents, the Modified GMP approach has divided the scope of the work into two main parts: the main contractor's direct works and Modified GMP works packages. For direct works, the traditional model of procurement is adopted. For the rest of the works (approximately 20% of the contract value), the GMP works packages are developed and the scope of work is defined. Those packages enable the contractor's expertise and innovation to be harnessed to construct in better and more efficient ways, such as ensuring higher levels of sustainability and construction efficiency.

Table 2.4 Partnering practices of the surveyed projects (II) (Chan et al., 2007a) (with permission from The Hong Kong Polytechnic University)

	Was partnering adopted?		At what stage was it decided to adopt partnering?				
	Yes	No	Feasibility	Outline design	Detailed design	Tender	Construction
Residential							
i. The Orchards	✓						✓
ii. Public Housing Development at Eastern Harbour Crossing Site Phase 4	✓						✓
Commercial							
i. 1063 King's Road	✓						✓
ii. Chater House	✓						✓
iii. Alexandra House Refurbishments	✓				✓		
iv. Tradeport Hong Kong Logistics Centre	✓				✓		
v. Landmark Redevelopment Phase 6 (York House)	✓		✓				
vi. Three Pacific Place	✓						✓
vii. TKO Technology Park	✓		✓				
viii. Wynn Resorts, Macau		✓					
ix. Omni Berkshire Place Hotel Renovation, New York		✓					
Infrastructural							
i. MTRC Tsim Sha Tsui Railway Station Modification Works	✓					✓	
ii. MTRC Contract 604 – Yau Tong Station	✓						✓
iii. MTRC Contract – Tung Chung Cable Car	✓					✓	

projects in Hong Kong. A selective tendering method is also adopted in most of the surveyed projects in order to ensure sufficient competition among the qualified contractors.

Negotiated tendering is also utilised in several surveyed GMP projects due to a long-term close working relationship between the employer and contractor.

In addition, as shown in Table 2.3, TCC and GMP can be adopted at various stages throughout the whole life of the project, but primarily at the outline design or detailed design stages. However, more recent projects have adopted TCC and GMP approaches at an earlier stage in order to draw on the builder's expertise and innovative ideas in project design, construction methods and selection of materials. For example, Hong Kong Land implemented a GMP contract in the York House Project at the feasibility stage to seek enhancements in buildability, innovation and efficiencies by acquiring the perceived benefits and alleviating unpredictable risks together with the partnering contractor.

The partnering practice adopted in the surveyed TCC and GMP projects is further revealed in Table 2.4. The majority of projects (85.7% of the total surveyed projects) utilised TCC and GMP with partnering in order to align individual goals to the common objectives of a project, facilitate communication among the project parties, and enhance the implementation of the gain-share/pain-share mechanism associated with TCC and GMP projects.

Most of the surveyed TCC and GMP projects adopted a partnering approach during construction. The infrastructure sector (i.e. the Mass Transit Railway Corporation) included a partnering provision in the contracts for the Tsim Sha Tsui Railway Station Modification Works and Tung Chung Cable Car Project at the tender stage. More recently, the partnering arrangement has been implemented earlier, at the feasibility stage, such as the York House and the TKO Technology Park, as long as the employer and the contractor have established a long-term close working relationship over a number of years.

Chapter summary

This chapter explored and discussed the history and development of TCC in the United Kingdom, Australia and Hong Kong. There were a plethora of successful large-scale TCC and GMP projects in urban areas. However, TCC is not a panacea, as the outcomes of some projects were not as desirable as expected. Hence, the benefits, difficulties and critical success factors will be analysed in the later chapters.

3 Perceived benefits of using target cost contracts

Introduction

The construction industry is regarded as a very competitive and risky business, as Chan et al. (2003) have stated. For projects procured with the traditional fixed-price lump-sum contract, limited trust among project stakeholders, and lack of incentives and common goals often lead to a confrontational working atmosphere, finally resulting in undesirable project performance (Construction Industry Review Committee, 2001; Walker and Hampson, 2003). Contractors are not motivated to devote their efforts beyond the minimum contractual requirements.

Some procurement methods have been introduced since the 1990s in order to meet the changing requirements of clients (Masterman, 2002). Previous overseas literature has revealed that TCC and GMP contracts can bring significant benefits to all contracting parties, if they can be properly structured, implemented and managed (Trench, 1991; Walker et al., 2000). In this chapter, the merits and rationale of TCC and GMP procurement strategies will be explored and discussed.

Motives for introducing TCC

An industry-wide empirical survey conducted by D.W.M. Chan et al. (2011b) determined the motives behind adopting TCC and GMP. As shown in Table 3.1, the most common motive for adopting TCC and GMP procurement strategies is 'to generate an incentive to achieve cost saving'. TCC involves a gain-share/pain-share mechanism that offers enormous incentives for contractors to minimise costs, innovate, construct effectively and overcome difficulties, according to Boukendour and Bah (2001). The second most important motive is 'to develop better working relationship'. TCC and GMP approaches promote a partnering spirit to gain deeper collaboration between the employer and the contractor. The adjudication committee included in the regular partnering review meetings under the TCC and GMP arrangements may help both parties to discuss any problems associated with the project and settle any confrontational issues (Chan et al., 2003).

Table 3.1 Frequency distribution of the motives for implementing TCC and GMP (Chan et al., 2007a) (with permission from The Hong Kong Polytechnic University)

Motive for TCC and GMP	All respondent group		Ranking
	Frequency	Percentage	
To generate an incentive to achieve cost saving	26	68.4%	1
To develop better working relationship	25	65.8%	2
To tap into contractor's expertise in design	24	63.2%	3
To set an agreed ceiling price at main contract award in case of GMP contracts	23	60.5%	4
To improve risk management and control	22	57.9%	5
Greater time saving by overlapping design and construction	15	39.5%	6
To enhance quality of constructed facilities	13	34.2%	7
Previous successful experience with TCC and GMP	11	29.0%	8
Need an 'open-book' accounting arrangement	8	21.1%	9
Total	38		

The third one is 'to tap into contractor's expertise in design'. Since the contractor can be involved earlier during design development, the construction works can be started before completing the entire design. Moreover, environmental issues and higher buildability factors are amalgamated into the design (Hong Kong Housing Authority, 2006).

'To set an agreed ceiling price at main contract award in case of GMP contracts' is the fourth most frequent motive. When compared with TCC and GMP, the fixed price stated in the traditional lump-sum contract may be different from the final price at completion, but GMP provides a price ceiling in order to reduce the cost variations for the employer (National Economic Development Office, 1982; Mills and Harris, 1995). Therefore, both parties' project revenues are foreseeable.

Apart from the aforementioned motives, 'to improve risk management and control' attracts both the client group and the consultant group. For example, the Mass Transit Railway Corporation, the major sole railway service provider in Hong Kong, implemented the first formal TCC for an underground railway station modification works project in order to align the project team's competence to the high risk profile (Avery, 2006) and to share the risks as pre-agreed between the employer and the builder (Mass Transit Railway Corporation, 2003).

More involvement of the client, an open-book accounting regime, confirmation of responsibility and quantification of risk in terms of costs are also the key characteristics of TCC and GMP (Boukendour and Bah, 2001; National Economic Development Office, 1982; Wong, 2006).

Table 3.2 Perceived benefits of TCC and GMP (Chan et al., 2007a) (with permission from The Hong Kong Polytechnic University)

Perceived benefits of TCC and GMP	HKHA (2006)	Tang (2005)	Cheng (2004)	Fan and Green-wood (2004)	Sadler (2004)	Tang and Lam (2003)	Bouke-ndour and Bab (2001)	Perry and Barnes (2000)	Patter-son (1999)	Gander and Hemsley (1997)	Chevin (1996)	Mills and Harris (1995)	Trench (1991)	NEDO (1982)	Total no. of hits for a certain benefit
Cost control															
Greater price certainty and better control of overspending	✓	✓	✓	✓	✓			✓	✓	✓		✓	✓	✓	10
Client provides financial incentives for contractor to achieve cost savings	✓	✓	✓	✓	✓	✓	✓	✓					✓		9
Risk sharing on cost overruns	✓				✓		✓	✓				✓	✓		6
Time control															
Fast track project by allowing early start of construction before the design is fully developed		✓											✓		2
More effort of client's involvement in problem solving					✓	✓							✓		3
Earlier settlement of final project account	✓				✓					✓					3

Benefits	1	2	3	4	5	6	7	8	9	10	11	12	13	14	Total
Greater flexibility of accommodating changes	✓	✓	✓		✓									✓	5
Quality control															
Greater client's control over building design and subcontracting process	✓	✓												✓	3
Selection of a right working team					✓									✓	2
Early contribution by contractor to both design and construction	✓	✓			✓								✓		4
Better estimate of the cost of quality work			✓		✓										2
Working relationship															
Incentives for effective collaboration between client and contractor	✓				✓	✓		✓				✓	✓		6
Conducive to improving partners' working relationship via partnering	✓				✓			✓							3
Total no. of benefits identified from each publication	8	7	5	2	9	3	3	5	1	2	2	2	6	3	58

Note: The previous studies are ranked in decreasing chronological order of year of publication followed by alphabetical order of authors. HKHA = Hong Kong Housing Authority; NEDO = National Economic Development Office.

Perceived Benefits of TCC

Literature review

TCC and GMP can benefit the project by controlling time, cost and quality, as well as improving the working relationship between project stakeholders. Chan et al. (2007a) summarised the perceived benefits of TCC and GMP from fourteen reports. The corresponding frequencies of their citations are shown in Table 3.2.

Cost control

TCC and GMP provide a more genuine price ceiling or target cost for the construction project, so they can limit the risk of cost uncertainty borne by employers (Patterson, 1999; Perry and Barnes, 2000). In particular, by utilising GMP, employers are only responsible for the agreed guaranteed maximum price in the contract. The amount can only be changed if additional work is needed and approved by the client. The contractor is required to solely bear the costs exceeding the GMP. The risk of going over budget is shifted to the contractor (Mills and Harris, 1995). The employer can thus prevent overspending due to this special characteristic of the TCC and GMP procurement strategies. Therefore, by limiting the provisions for modifying the contract price and adopting a strict variation procedure, the TCC and GMP approaches protect the client from being exposed to substantial cost increases (Lewis, 1999).

In addition, TCC and GMP are regarded as a gain-share/pain-share mechanism which awards contractors for the costs saved, but penalises them for the exceeded costs. Therefore, the contractor has a high incentive to work efficiently and to improve cost savings (Fan and Greenwood, 2004; Boukendour and Bah, 2001). Furthermore, TCC and GMP motivate the client and contractor to co-operate in order to minimise costs, since they can both gain from the cost savings (Tang and Lam, 2003). According to Perry and Barnes (2000), if the contractors' share of the savings percentage is increased, they will be motivated to ensure the certainty of yield under TCC.

Meng and Gallagher (2012) analysed different procurement approaches with respect to cost certainty. As shown in Table 3.3, TCC performed favourably in terms of 'cost certainty' and was ranked second. The surveyed TCC projects

Table 3.3 Cost certainty under different payment methods (Meng and Gallagher, 2012) (with permission from Elsevier)

	Fixed-price	TCC	Payment based on final outcomes	Cost plus fee
No. of projects	20	17	12	11
Under budget	7 (35.0%)	5 (29.4%)	1 (8.3%)	1 (9.1%)
On budget	7 (35.0%)	4 (23.5%)	2 (16.7%)	1 (9.1%)
Over budget	6 (30.0%)	8 (47.1%)	9 (75.0%)	9 (81.8%)

were 29.4% under budget, 23.5% on budget and 47.1% over budget. Over 50% of TCC projects were within budget (either under budget or on budget), which was mainly attributable to the contractor's incentives to share cost savings, or at least to ensure the target fee. Nevertheless, loss sharing may lead to less pressure on builders' cost controls. Therefore, TCC may lead to a higher likelihood of going over budget when compared with fixed-price contracts.

Time control

Since a typical construction project under the traditional design-bid-build procurement method requires a sequential approach to the design and construction process, it cannot allow for fast-track arrangements. On the contrary, TCC and GMP enable the early commencement of construction activities prior to completion of the design (Frampton, 2003). As design overlaps with construction substantially, similar to a design-and-build contract, the duration of overall project development can be shortened. Furthermore, according to Tang and Lam (2003), when compared with traditional contracts, clients can be involved in problem solving, so decisions on any variations can be made more effectively and efficiently. The TCC and GMP procurement methods may thus accelerate the problem solving process (Trench, 1991).

In addition, the arrangements for changes with the TCC and GMP procurement methods are pre-agreed by the employer and the builder, which helps to mitigate the occurrence of claims/disputes and allows earlier preparation and agreement of the final project account when compared with conventional procurement methods (Gander and Hemsley, 1997). Besides, the adjudicating mechanism enhances efficiency by settling the final project account earlier, whereas traditional contracts usually instigate protracted debates on variations, leading to project delays. Moreover, TCC and GMP approaches enable greater flexibility in accommodating design variations due to the straightforward variation claiming mechanism as well as the open-book accounting arrangements (Mills and Harris, 1995). In comparison with traditional procurement strategies, managing changes under TCC and GMP schemes can thus be less time-consuming and more explicit.

Quality control

Adopting TCC and GMP reaps benefits in construction quality. As mentioned by Cheng (2004), traditional contracts over-emphasise saving costs and sacrifice quality. In sharp contrast, since TCC and GMP lay down a reasonable target price, they facilitate the tendering process so that the domestic subcontractors' works packages can be tendered on an open basis. This enables the client to acquire competitively priced tenders from eligible subcontractors and specialists (Tay et al., 2000). Thus, TCC and GMP can help the employer to choose a suitable project team with sufficient experience and ability to advance the client's design intent (Trench, 1991). Moreover, these methods eliminate non-value-adding multi-tier subcontracting and maintain the quality standards of built facilities and workmanship. In addition, TCC and GMP increase the overall quality as, during the pre-contract and post-contract stages, the employer can have greater control over the design. Thus, the final design complies with the original design intent described in the employer's project brief (Hong Kong Housing Authority, 2006). Moreover, the employer can help to resolve any problems (Tang and Lam, 2003). Apart from the employer, since contractors are involved at the design stage, they can make improvements in terms of construction cost, design, materials, project programming, alternative construction techniques and other buildability-related components (Hong Kong Housing Authority, 2006). TCC and GMP also enable better estimation of the cost of construction projects that satisfies the client's demands in terms of quality (National Economic Development Office, 1982). Furthermore, due to the early involvement of the contractor during the design stage, TCC and GMP can come up with a more cost-effective procurement strategy and a more buildable design, resulting in higher value added to the project by the contractor, for instance by encouraging better integration of building reinforced concrete construction and services installation, increasing innovations throughout the entire project life and utilising the contractor's expertise in specialist design and prefabrication techniques.

Working relationship

According to Bower et al. (2002), TCC and GMP contracting strategies can motivate contractors to achieve better project performance and higher value by aligning their individual financial goals with the common objectives of the project.

In particular, the gain-share/pain-share initiative promotes efficient collaboration between the employer and builder in order to reduce the outturn cost (Chevin, 1996; Sadler, 2004). Due to the involvement of all related project stakeholders, design development before construction can minimise conflicts and disputes at the later stages. TCC also allows the determination of suitable ownership of risks between client and contractor, and encourages the contracting parties to consent to an equitable allocation of risks, resulting in the client's long-term best benefit (Sadler, 2004). Furthermore, more communication opportunities and a fair dispute resolution mechanism during adjudication meetings lead to fewer disputes/claims, help to improve adverse working relationships between project team members and increase inter-disciplinary efforts to benefit the project (Ting, 2006).

In addition, TCC and GMP are conducive to fostering a 'partnering' spirit in the relationships among the different contracting parties, involving the client, main contractor, subcontractors and consultants, with the common goal of bringing a more co-operative and less litigious philosophy to the project (Tang and Lam, 2003; Hong Kong Housing Authority, 2006). Chan et al. (2004) stated that GMP procurement in numerous building projects and the incentivisation agreement adopted in railway infrastructure projects in Hong Kong have shown that it can nurture a more co-operative working relationship and a gain-share/pain-share working culture, mainly as a result of the perceived 'partnering' concepts among all contracting parties.

Table 3.4 Ranking of the perceived benefits of TCC and GMP in Hong Kong (all respondents) (D.W.M. Chan et al., 2011a) (with permission from Elsevier)

Benefits of TCC and GMP	All respondent group	
	Mean	Rank
Early settlement of final project account	4.22	1
Conducive to improving partners' working relationship via the gain-share/pain-share mechanism and partnering arrangement	4.11	2
Bring in expertise in building design and innovations in construction methods and materials from contractor to enhance the buildability of the project	4.11	2
Client provides financial incentives for contractor to achieve cost savings	3.97	4
Fast track project by allowing early start of construction before the design is fully developed	3.92	5
Achieve better value for money	3.92	5
The gain-share arrangement helps establish mutual objectives and produce an integrated, trustful working team	3.86	7
Provide guarantee of avoiding budget overrun at main contract award for the client in case of GMP contracts	3.84	8
Early award of contract can allow advanced works packages (e.g. demolition, foundation, etc.) to be included in GMP or target cost	3.84	8
More effort of client's involvement in problem-solving and subcontractor selection	3.81	10
More opportunities for participants to express opinions and concerns openly and freely	3.81	10
Limit the entitlements for claiming variations by contractor	3.73	12
Enable a more equitable risk apportionment among project participants	3.73	12
Domestic subcontractor's works packages are competitively tendered by approved or pre-qualified subcontractors and specialists on an open-book basis after the award of TCC and GMP contract as design develops	3.68	14
Provide a dispute resolution mechanism via adjudication committee, leading to reduction in disputes	3.57	15
Greater client's control over design consultants, main contractor and subcontractor	3.41	16
Contractor takes all the risks in design development via TCC and GMP allowance in the tender	3.30	17
Number (N)	37	

Note: Items were rated on a five-point Likert scale, with 1 = strongly disagree, 3 = neutral and 5 = strongly agree.

Findings from a questionnaire survey

Table 3.4 shows the relative significance of the perceived benefits of TCC and GMP which were identified by the questionnaire survey conducted by D.W.M. Chan et al. (2011b).

'Early settlement of final project account' is the most significant benefit under TCC and GMP. It echoes the finding of Gander and Hemsley (1997) that, when compared with traditional fixed-price contracts, preparation and agreement of the final project account with TCC and GMP tend to be accomplished earlier, as both the time and cost implications of any potential variations have been pre-agreed between the employer and the contractor. Therefore, this may help to reduce potential claims and intractable disputes throughout the construction process.

TCC and GMP are also beneficial to the project by integrating the contractor's innovative ideas and expertise during the design and construction stages to increase the constructability of the original design, because under the TCC and GMP approaches, the contractor can be involved in the early design process and provide technical advice or recommendations on buildability and environmental issues, which helps to develop a cost-effective design (Wong et al., 2006).

Regarding the second major benefit, 'conducive to improving partners' working relationship via partnering' can be achieved through the gain-share/pain-share mechanism with the mutual objective of maximising cost savings through the partnering arrangement (Chan et al., 2003). The conventional working relationships between different project stakeholders are usually confrontational, often leading to contractual claims and even litigation. Moreover, not only can the financial incentives under TCC and GMP contracts motivate the contractor, they can also help to achieve alignment of project objectives for different industry stakeholders. Ting (2006) believed that the incentivisation approach could foster a more proactive and co-operative working relationship between various contracting partners and encourage the cultural change from the conventional antagonistic approach to collaborative contracting.

TCC and GMP are conducive to instilling a 'partnering spirit' in the working culture among contracting parties (Tang and Lam, 2003; Hong Kong Housing Authority, 2006). This echoes another two benefits: 'client provides financial incentives for contractor to achieve cost saving' and 'the gain-share arrangement helps establish mutual objectives and produce an integrated, trustful working team'.

There are another two merits: 'achieve better value for money' and 'more effort of client's involvement in problem solving and subcontractor selection'. If TCC and GMP can generate competitive pricing and incentives for innovation, they can motivate contractors to improve project performance and achieve better value (Construction Industry Review Committee, 2001). Sadler (2004) suggested that having fewer scope changes/variations can ensure that TCC and GMP contracts are administered as initially intended and help the client to gain greater value for money. The employer is also required to be involved and commit to the contract as stated in the tendering assessment and project management strategy. Hence, this kind of procurement strategy is advantageous for construction projects (Tang and Lam, 2003; Sadler, 2004).

In particular, regarding the benefit 'client provides financial incentives for contractor to achieve cost saving', only the respondents from clients and consultants considered it to be the top-ranking benefit. However, it was not regarded as an important benefit from the contractor's perspective, ranking very low. One of the apparent differences between lump-sum contracts and TCC (or GMP contracts) is which party may gain from the savings.

The contractor under a lump-sum contract may gain the entire savings, whereas under TCC and GMP contracts, the savings are shared between client and contractor. Therefore, clients favour TCC and GMP contracts over fixed-price contracts, whereas the contractors prefer fixed-price contracts rather than TCC and GMP arrangements. Furthermore, this ranking difference may be attributed to the different expectations and interpretations of the rationale of TCC and GMP regarding financial incentives between the two parties. According to Boukendour and Bah (2001), the clients, as well as their consultants, may believe the gain-share approach to be a strong incentive to achieve cost savings.

Nevertheless, it is hard to achieve cost savings when there are ambiguous project scope of work and unexpected risks related with TCC and GMP procurements (Fan and Greenwood, 2004), such as those resulting from deficient design details during the tender stage. Generally speaking, Meng and Gallagher (2012) found that the incentives play a driving role in achieving project success and encouraging the most suitable practice for implementation. They also perceived that incentives could help align the individual objectives of various project stakeholders and that multiple incentives might improve overall project performance, while a single incentive could lead to a greater focus on a certain performance area.

Although 'limit the entitlements for claiming variations by contractor' was ranked low (twelfth) as a whole, both contractors and consultants perceived it as an important benefit, since various potential changes are clearly defined and pre-agreed by the contracting parties when adopting TCC and GMP contracts, as supported by Gander and Hemsley (1997). Thus, the contractor may be more willing to accommodate variations of design in order to gain cost savings (Mills and Harris, 1995). Nevertheless, clients expressed the opposite view, since they may still be required to accept the project changes and additional works lodged by the builders (Fan and Greenwood, 2004), especially when the target cost or GMP decided early at the design stage are not agreed thoroughly. These opposing views may be due to the different areas and levels of participation in the construction project. As builders and consultants usually manage claims for variations, they may find that the claims tend to be lowered or eliminated when implementing TCC and GMP forms of procurement. Hence, they perceived that claim occurrence is reduced under TCC and GMP approaches.

Not all parties agreed that 'conducive to improving partners' working relationship via the gain-share/pain-share mechanism and partnering arrangement' was a major benefit. The consultants were less impressed by this benefit than clients and contractors. In the initial partnering workshop, both employer and contractor achieve a mutual target – saving cost, by cultivating a more harmonious and co-operative working relationship throughout the construction project (Chevin, 1996; Chan et al., 2004; Sadler, 2004), while the savings may not be shared by the team of consultants.

Chapter summary

All in all, the perceived benefits of TCC and GMP include the following:

- They provide a guarantee of avoiding going over budget at main contract award for the employer in the case of GMP contracts.
- They enable the client to offer financial incentives to the contractor to achieve cost savings.
- They allow early commencement of construction before finalising the design.
- They utilise contractors' expertise in innovations and building designs to give advice on construction methods and materials in order to improve the buildability of the project.
- They provide a dispute resolution mechanism by means of the adjudication committee, resulting in a reduction in disputes or claims.
- They allow the final project account to be settled earlier, since the valuation of variations is agreed progressively at the construction stage.
- They are conducive to enhancing partners' working relationship through the partnering arrangement.
- They constrain the contractor's entitlements to claim for variations.
- They encourage a more equitable risk apportionment between project participants.
- They provide opportunities for participants to express views and concerns openly and freely.
- They help to develop mutual goals and establish an integrated, trustful working team.

4 Potential difficulties in implementing target cost contracts

Introduction

TCC and GMP procurement methods are still not very common in the Hong Kong construction industry. It is necessary to investigate the potential difficulties when adopting TCC and GMP procurement strategies, as any lessons learnt in Hong Kong could be of global interest. This chapter aims to identify potential difficulties with the current practices of TCC and GMP, and to provide useful insights for improvement by alleviating the effects of risk factors and the occurrence of potential difficulties encountered in future TCC and GMP projects.

Potential difficulties of TCC

Literature review

A desktop review of the literature conducted by Chan et al. (2007a) indicated that some typical difficulties do exist when carrying out the TCC and GMP approaches in spite of the perceived benefits when implementing TCC and GMP concepts. Table 4.1 lists the common difficulties identified from related publications together with the associated frequencies of their citations.

Unclear definition of changes in scope of work

Unclear definition of changes in scope of work is the primary problem faced when adopting the TCC and GMP approaches (Gander and Hemsley, 1997). Arguments may result from unclear explanations of any scope changes, which tend to lead to conflict between the client and contractor in attempting to achieve their intentions (Cheng, 2004; Fan and Greenwood, 2004). The builder tends to regard changes as 'scope changes' in order to secure the greatest chances of additional payment. However, the client would like to class differences under 'design development' in order to reduce excessive costs, not to mention the wish to achieve maximum cost savings. There is a difficulty in assessing the revised contract price when the contractor proposes an alternative design and extra time is needed to evaluate the cost

Table 4.1 Potential difficulties in implementing TCC and GMP (Chan et al., 2007a) (with permission from The Hong Kong Polytechnic University)

Potential difficulties in implementing TCC and GMP	HKHA (2006)	Tang (2005)	Cheng (2004)	Fan and Greenwood (2004)	Sadler (2004)	Tang and Lam (2003)	Perry and Barnes (2000)	Tay et al.' (2000)	Patterson (1999)	Minogue (1998)	Gander and Hemsley (1997)	Chevin (1996)	Mills and Harris (1995)	Total no. of hits for a certain difficulty
Unclear definition of changes in scope of work, leading to unnecessary disputes	✓	✓	✓	✓	✓			✓	✓		✓	✓		9
Difficult to evaluate the revised contract price when an alternative design is proposed by the contractor					✓	✓	✓							3
Higher costs to adopt TCC and GMP for contractor to cover additional risks		✓	✓		✓	✓		✓		✓	✓		✓	8
Increased commitment and involvement by project managers and design consultants in evaluating tenders for domestic subcontracts	✓				✓									2
Design development must keep pace with contractor's programme for tendering domestic subcontractor's works packages, otherwise delays may occur	✓													1
Unfamiliarity with or misunderstanding of TCC and GMP concepts		✓			✓								✓	3
Too complicated a form of contractual agreement								✓						1
Total no. of difficulties identified from each publication	3	3	2	1	5	2	1	3	1	1	2	1	2	27

Note: The previous studies are ranked in decreasing chronological order of year of publication followed by alphabetical order of authors. HKHA = Hong Kong Housing Authority.

implications again (Tang and Lam, 2003). Moreover, Tay et al. (2000) stated that it is extremely hard to manage, as this is a feature of the TCC and GMP approaches. Thus, Trench (1991) considered that, with numerous variations expected, the TCC and GMP procurement forms might not be a suitable strategy for contracts where it is difficult to define the scope of work.

Higher cost premium for TCC and GMP

Generally speaking, more responsibilities are borne by the contractor under TCC and GMP contracts than traditional ones, including an allowance for design development and unexpected risks (Sadler, 2004). Lewis (2002) suggested that the main contractor could shift the risks to subcontractors. It was also suggested that the bid price be increased for the contractor to fulfil the price guaranteed, with additional risks included. Under most circumstances, tender sums under GMP projects may be around 1–3% greater than equivalent tenders sought under a JCT 80 with quantities standard form of contract in favourable conditions, and the contract sum is the de facto highest price assured (Mills and Harris, 1995). That is to say, an amount of cost certainty is obtained by the client, yet the price is generally not the lowest. However, TCC and GMP contracts can be a desirable choice when a fixed price is more crucial than ensuring the lowest price.

Greater commitment by project participants

A higher level of commitment and involvement are required for the TCC and GMP approaches from all project stakeholders involved in the contract according to the methodology of tendering; this not only applies to the main TCC or GMP, but also to the domestic subcontractor's works packages on an individual basis (Tang and Lam, 2003). Sadler (2004) felt that the employer has to monitor tightly and participate more in a project when adopting the TCC and GMP approaches, since the contractor has committed to the ceiling price before developing the design. Delays may result if the design development does not keep pace with contractor's programme for tendering the domestic subcontractor's works packages. More personnel may be involved in the project due to these extra administrative requirements. Moreover, higher fees may be required by design consultants when assessing tenders for domestic subcontracts after awarding the main contract (Hong Kong Housing Authority, 2006).

Unfamiliarity with TCC and GMP methodology

The TCC and GMP contracting strategies are relatively new concepts in the Hong Kong construction industry. Cheng (2004) stated that disputes are easily generated between the parties because project stakeholders are not familiar with the corresponding contractual arrangements. It is difficult to set allowances for design development and unforeseen risks, and to determine the cost-sharing formula for TCC and GMP schemes. A greater potential for drafting mistakes and misinterpretation of liabilities between the parties may result from the lack of standard forms of TCC and GMP contracts (Gander and Hemsley, 1997). More administrative effort and support are needed to establish and carry out the complicated form of contractual agreement, which is sometimes not feasible in certain projects (Sadler, 2004).

Findings from a questionnaire survey

Table 4.2 shows the general ranking of the difficulties of TCC and GMP in Hong Kong resulting from an industry-wide empirical questionnaire survey (Chan et al., 2010b). The two major problems with TCC and GMP were 'design development must keep pace with main contractor's programme for tendering the domestic subcontractor's works packages' and 'clients had to be more involved in a project'.

Delays may occur if design development progress fails to keep pace with the main contractor's programme for tendering the domestic subcontractor's

Table 4.2 Ranking of the potential difficulties of TCC and GMP in Hong Kong (all respondents) (Chan et al., 2010b) (with permission from American Society of Civil Engineers)

Difficulties in implementing TCC and GMP	All respondent group	
	Mean	*Rank*
Design development must keep pace with main contractor's programme for tendering the domestic subcontractor's works packages, otherwise delays may result	4.03	1
Clients had to be more involved in a project	3.97	2
Disputes over whether architects'/engineers' instructions constituted TCC and GMP variations or were deemed to be design development, i.e. unclear scope of work	3.63	3
Lack of standard form of TCC and GMP contract in the local context	3.57	4
Not suitable for projects where it is difficult to define the scope of work early	3.43	5
Increased commitment and involvement by project managers and design consultants in evaluating tenders (mainly on technical elements) for domestic subcontracts after the award of main contract, i.e. potential for incurring higher consultant fees	3.40	6
Longer time in preparing contract documents	3.17	7
Unfamiliarity with or misunderstanding of TCC and GMP concepts by senior management	3.17	7
A project team may find it difficult to adapt to this new way of working	2.91	9
Too complicated a form of contractual agreement	2.57	10
Difficult to develop trust and understanding from contractor as a project team	2.49	11
Difficult to launch subcontracting with back-to-back contract terms	2.49	11
Number (N)	**35**	

Note: Items were rated on a five-point Likert scale, with 1 = strongly disagree, 3 = neutral and 5 = strongly agree.

works packages (Hong Kong Housing Authority, 2006). Intractable arguments may result from ambiguous explanations of any scope changes, which tend to lead to conflict between client and the contractor in their efforts to achieve their own intentions (Cheng, 2004; Fan and Greenwood, 2004).

Furthermore, the employer has to monitor closely and participate more in the project when adopting TCC and GMP approaches, since the contractor has committed to the ceiling price before developing the design (Sadler, 2004). A higher level of commitment and involvement is required for TCC and GMP approaches from all project contract stakeholders based on the methodology of tendering; this not only applies to the main TCC or GMP, but also to the domestic subcontractor's works packages on an individual basis (Tang and Lam, 2003).

Furthermore, there is also the third most crucial hindrance when handling TCC and GMP projects: 'disputes over whether architects'/engineers' instructions constituted TCC and GMP variations or were deemed to be design development', as in the interview findings of Chan et al. (2007a). Design development changes under the conditions of TCC and GMP schemes will not trigger an adjustment of the GMP value or target cost because they are included in the unchanged lump-sum price of the main contractor's direct works. However, alterations in the scope of work can lead to changes in target cost or GMP value (Fan and Greenwood, 2004; Hong Kong Housing Authority, 2006). The contractor tends to regard any change as a 'scope change' in order to secure the best prospects of securing additional payment as a result of unclear scope of work. However, the client would prefer to class differences as 'design development' in order to reduce cost increases. This may be the greatest potential area for conflict, particularly if the target cost is ascertained at the early design stage (Tay et al., 2000). In addition, it can be difficult to define the degree of design development changes. Through the adjudication committee's description of the status of a variation submission and determination of the categorisation of any variations the contractor may submit, the development of an adjudication process with higher effectiveness and efficiency is critical for TCC and GMP projects (Hong Kong Housing Authority, 2006). It is also important to arrive at agreement by common consent over the evaluation of changes as quickly as possible so that the progress of the scheme is not disturbed.

Before the decision on whether to use the TCC and GMP forms of procurement strategy, decision makers have to focus particularly on the difficulties encountered: (1) 'lack of standard form of TCC and GMP contract in the local context' (Gander and Hemsley, 1997), though this has been resolved since 2005 due to the launch of NEC3; (2) 'not suitable for projects where it is difficult to define the scope of work early' (Chevin, 1996); (3) 'increased commitment and involvement by project managers and design consultants in evaluating tenders for domestic subcontracts after the award of main contract leading to the potential for incurring higher consultant fees' (Tang and Lam, 2003); (4) 'longer time in preparing contract documents';

and (5) 'unfamiliarity with or misunderstanding of TCC and GMP concepts by senior management' (Chan et al. 2007).

With regard to the profound difficulties encountered, as shown in Table 4.2, TCC and GMP contracts are vulnerable when there is a need for the design development to keep pace with the tendering of the domestic subcontractor's works packages by the main contractor, and when the client is required to be heavily involved in the whole project delivery process. This echoes the findings reported by Sadler (2004) and the Hong Kong Housing Authority (2006).

In respect of 'increased commitment and involvement by project managers and design consultants in evaluating tenders for domestic subcontracts, i.e. potential for incurring higher consultant fees' (ranked sixth as a whole), the client may not find it difficult (ranked ninth) as the tender evaluation of subcontracts is mainly conducted by the team of consultants as well as the main contractor's project manager, not by the client.

Chapter summary

The following difficulties may arise during the implementation of TCC and GMP schemes:

- Arguments may arise about whether architects'/engineers' instructions are regarded as TCC and GMP variations or are deemed to be design development – unclear scope of work.
- Project managers and design consultants bear greater commitment and involvement when assessing tenders for domestic subcontracts after awarding the main contract – potential for incurring consultants' fees.
- Design development must keep pace with the contractor's programme for tendering the domestic subcontractor's works packages, otherwise there may be delays.
- More time is required when preparing contract documents and assessing the revised contract price for an alternative design proposed by the builder.
- Senior management may be unfamiliar with or misunderstand TCC and GMP concepts.
- It may be difficult to develop mutual trust and understanding between the contractor and the rest of the project team.
- The form of contractual agreement may be over-complicated.
- There is a need for a greater degree of involvement of clients in the project.
- It can be hard for the project team to adapt to a new way of working (e.g. joint efforts between consultants and main contractor in design work).
- They may be unfavourable for projects where it is difficult to define the scope of work early on.
- It can be difficult to launch subcontracting with back-to-back contract terms.

5   Critical success factors for adopting target cost contracts

Introduction

There are a variety of problems that beset the construction industry, including misalignment of objectives among project stakeholders, unfair risk allocation, insufficient incentives to improve project performance and limited mutual trust, resulting in project delays, cost overruns, difficulty in resolving claims and a win–lose atmosphere (Moore et al., 1992; Chan et al., 2004).

Many countries, including the United States, United Kingdom and Australia, place emphasis on TCC and GMP contracting strategies. Two pilot projects in the United Kingdom achieved 8–10% cost reduction, 5–20% faster programme completion and 90–95% rework reduction (Nicolini et al., 2001). However, Harris (2002) found that the New Wembley National Stadium located in London, which adopted the GMP method, cost £757 million more than the initial estimated budget of £200 million in 1996, and was completed two years behind schedule due to many reasons, such as two litigation cases between the main contractor and the steel contractor.

This chapter thus focuses on identifying and analysing the critical success factors when utilising TCC and GMP contracts by evaluating empirically the views of local construction practitioners on which elements bring success to TCC and GMP projects. It is worthwhile exploring the essential success factors in order to achieve future construction excellence.

Critical success factors for TCC

Literature review

Chan et al. (2007a) reviewed and summarised the critical success factors driving the success of projects under TCC and GMP schemes, as shown in Figure 5.1.

Well-defined scope of work

According to Tang (2005), the scope of the builder's work should be well defined in the client's project brief to avoid disputes between the employer

Figure 5.1 Critical success factors for TCC and GMP projects (Chan et al., 2007a)
(with permission from The Hong Kong Polytechnic University).

and the builder over whether the employer's changes relate to scope
variation or design development. This also requires that both client and contractor understand the TCC approach and the roles and relationships under
this form of procurement (National Economic Development Office, 1982).
Moreover, Sadler (2004) stated that changes of scope or variations should
be kept to a minimum in order to administer TCC and GMP contracts as
intended and to increase the value for money of construction.

Demonstrated partnering spirit

Chan et al. (2004a) concluded that more effective communication and
harmonious working relationships among contracting parties could lead to
more satisfactory project outcomes. The success of TCC also requires a genuine
willingness to co-operate or demonstrate a partnering spirit within the project
team (Tay et al., 2000). After achieving the common goals and generating a
teamwork culture, disputes are always resolved. Furthermore, the project can
avoid tedious contractual claims which may require litigation for resolution.

It is therefore essential to establish an adjudication committee under the
TCC and GMP arrangements to manage the prompt resolution of disputes.
In addition, an open-minded attitude towards different parties' opinions is
of paramount significance (Tang and Lam, 2003). As advocated by Sadler
(2004), TCC relies greatly on mutual trust and fairness.

Right selection of project team

Suitable selection of the project team is important in developing mutual trust, effective communication, efficient co-ordination and productive conflict reconciliation (Chan et al., 2002). Hence, Gander and Hemsley (1997) recommended that experienced project parties are indispensable to achieving a successful TCC project, since inexperienced TCC and GMP contractors may lack clarity concerning their responsibilities. The National Economic Development Office (1982) also claimed that managerial efforts devoted by various project parties to formulating and administrating the contract are closely related to the success of the contract. The selection of a strong and experienced on-site contract management team is essential in monitoring and managing the operations of the TCC and GMP schemes from the beginning of the project.

Reasonable share of risks and cost saving

Mills and Harris (1995) stated that a clear and equitable sharing of risks between employer and contractor could lead to the success of a TCC or GMP project. Perry and Barnes (2000) suggested that clients should comprehend the importance of realistic target cost estimates, including appropriate risk contingencies. Sadler (2004) also advised that the client should be careful to manage a combination of fee and share which would not motivate the contractor to minimise the cost. However, a high sharing ratio for the contractor may motivate it to maximise the upward adjustment of the target cost that may occur during the course of the contract. Therefore, an appropriate allocation of cost savings is inextricably linked with the success of TCC and GMP contracts. Tang and Lam (2003) recommended that the shared percentage of cost savings for Hong Kong GMP construction projects should rely on the level of cost savings achieved, as indicated in Table 5.1.

Fan and Greenwood (2004) also advocated that contractors should comprehend the risks they are bearing, beware of undescribed work and of 'design development', and also ensure that the bids from their subcontractors indicate the risks that they will be bearing. Moreover, a decision has

Table 5.1 Shared saving percentage apportionment (Tang and Lam, 2003) (with permission from *Hong Kong Engineer*)

Scenario	Client's share	Contractor's share
Final out-turn cost > final GMP	0%	100%
Final out-turn cost < final GMP		
(a) Saving < 5%	67%	33%
(b) Saving = 5–10%	50%	50%
(c) Saving > 10%	33%	67%

to be made regarding the nature of the contracting procedures, the client's requirements and whether the work will ensure that the contractor is clear enough about what is actually involved within the scope of work so that a realistic assessment of the price can be made (Lewis, 2002).

Early involvement of the contractor in design development

It is crucial to take advantage of the competence of the contractor and suppliers at the design stage and before finalising the design (Sadler, 2004). As mentioned above, technical advice concerning buildability and environmental matters can thus be integrated into the design made by the contractor. The Hong Kong Housing Authority (2006) found that contractors' early involvement and advice concerning design, construction methods and materials are prerequisites for improving time, cost and quality performance.

Findings from a questionnaire survey

An empirical questionnaire survey conducted by Chan et al. (2010c) derived the following results with cross-reference to the published literature in order to complement each other for validation, if suitable.

Table 5.2 Ranking of the critical success factors for TCC and GMP (Chan et al., 2010c)

No.	Critical success factors for TCC and GMP	All respondent group	
		Mean	Rank
5	Reasonable share of cost saving and fair risk allocation	4.59	1
6	Partnering spirit from all contracting parties	4.51	2
4	A right selection of project team	4.46	3
2	Well-defined scope of work in client's project brief	4.43	4
9	Proactive main contractor throughout the TCC and GMP processes	4.41	5
7	Early involvement of the contractor in design development	4.35	6
3	Familiarity with and experience of TCC and GMP methodology among client, consultants, main contractor and subcontractors	4.16	7
10	Open-book accounting regime as provided by the main contractor in support of its tender pricing	4.08	8
8	Establishment of adjudication committee and meeting	3.78	9
1	Standard form of contract for TCC and GMP projects	3.41	10
	Number (N)	37	

Note: Items were rated on a five-point Likert scale, with 1 = strongly disagree, 3 = neutral and 5 = strongly agree.

Table 5.2 summarises the ranking of the critical success factors when implementing TCC as rated by the survey respondents. 'Reasonable share of cost saving and fair risk allocation' and 'partnering spirit from all contracting parties' were considered as the top two critical success factors for TCC and GMP contractual arrangements.

TCC and GMP provisions imply that builders bear more financial risks as they solely bear the costs exceeding the contract target cost as a result of uncertainties at the design stage (Stukhart, 1984). Therefore, Sadler (2004) emphasised that construction projects under TCC rely greatly on fairness and mutual trust. In respect of the unclear scope of work during tender stage under the TCC and GMP methodology, Mills and Harris (1995) discovered that establishing a reasonable target cost and sharing ratio of cost saving/loss between employer and contractor is crucial to the operation of TCC and GMP arrangements. Sadler (2004) also suggested that employers should ensure that the combination of target cost fee and share of the risks allocate the risks on an equitable basis, as well as the incentive to motivate the contractor. Perry and Barnes (2000) mentioned a strong case for avoiding establishing a contractor's sharing ratio of less than 50%. Moreover, Mills and Harris (1995) suggested that a clear and equitable allocation of risks between employer and contractor is also a prerequisite for the success of a construction project under TCC and GMP. Complicated or unfair risk apportionment can lead to intractable disputes and costly claims. Hence, Sadler (2004) suggested that the employer should allow the contractor to include an allowance for design development and potential risks clearly and reasonably within the tender. Fan and Greenwood (2004) recommended that contractors should be well prepared to realise the risks they will bear, pay attention to undescribed work at the 'design development' stage, and ascertain that their subcontractors' bids rationally show the potential risks they will be taking. The nature of contracting procedures, the client's requirements and the scope of work should be decided to generate a realistic assessment of the tender price (Lewis, 1999).

The study conducted by Tay et al. (2000) shows results consistent with the survey findings that partnering spirit or a close business relationship among all contracting parties is an essential element to achieve success of a TCC project. Chan et al. (2007a) conducted an interview with some industrial practitioners who also agreed that that partnering spirit needs to be developed hand-in-hand with TCC and GMP schemes to drive the success of a project. Partnering can enormously facilitate communication flows, build up mutual trust, resolve disputes and ameliorate the working relationship between the client's team and the contractor's team (Chan et al., 2004). Transparency of information exchange should increase parties' confidence and result in effective collaboration through the closer alignment of motivation. This partnership doctrine and open-minded attitude towards the opinions raised by different parties become particularly imperative for TCC and GMP contracts, since uncertain scope of work often exists at the initial phase of a project, and meanwhile the project team may be unfamiliar with the procurement process.

The Hong Kong Housing Authority (2011) promoted a partnering initiative for 'Quality Housing'. The objectives of this partnering approach were to improve team relationships, develop mutual trust, reinforce team communications and encourage a joint problem solving mindset. As the realisation of these goals relied on various stakeholders taking pride in their responsibilities and a sense of ownership of the ultimate product, the Hong Kong Housing Authority (2011) implemented the following ways to strengthen the commitment at two levels: (1) in order to pledge a commitment to construct quality buildings and achieve construction safety through partnering, a Partnering Charter was signed, and (2) since November 2002, partnering workshops have been held for the project stakeholders of all piling and building contracts.

The following are the typical partnering mechanisms in a construction contract (Hong Kong Housing Authority, 2011):

- At the initial stage, an external independent facilitator conducts start-up workshops. During the course of the project, regular partnering meetings are needed for relevant stakeholders to evaluate and manage the project performance regularly and identify priorities for enhancement.
- At project completion, an external independent facilitator holds close-out workshops.

Hong Kong Housing Authority (2011) further stated some partnering values:

- better communications and understanding among different contracting parties;
- improved working relationships, co-operation and mutual trust;
- better quality of work and workmanship;
- prompt addressing of site problems;
- smoother and more timely work progress;
- less paperwork;
- partners might be more proactive and willing to contribute to achieving mutual project objectives.

In order to cultivate a partnering spirit within the project team, openness of project accounting and a suitable adjudication arrangement are vital. The contractor's tender pricing should be accessible for investigation by the employer and its team of consultants with an appropriate auditing system. The National Economic Development Office (1982) also maintained that the utilisation of open-book accounting can lead to improved responsibility and a trustworthy working relationship. In addition, an adjudication process by an independent body is indispensable to resolve disputes promptly at site level and sustain an amicable working relationship. One of the previous interviewees further recommended that the adjudication process should be respected to avoid potential disputes (Chan et al., 2007a). The adjudication committee thus takes on a crucial role in preventing intractable disputes but the committee's success relies heavily on mutual trust, along with a partnering commitment between the teams of the client and contractor (Sadler, 2004).

Moreover, 'a right selection of project team' (ranked third) and 'proactive main contractor throughout the TCC and GMP processes' (ranked fifth) were regarded as another two important success factors for TCC and GMP projects. Gander and Hemsley (1997) echoed that the selection of an experienced project team is essential to successful TCC and GMP projects, because inexperienced or claim-conscious TCC and GMP contractors may lack clarity concerning their roles and responsibilities. Therefore, the contractor's initiative and competence to provide alternatives for best-value outcomes, mutual trust, and receptiveness and expertise for innovation by the project team and the contractor determine the success of TCC and GMP contracts (Gander and Hemsley, 1997). Recruitment of suitable partners with essential commitment and expertise, efficient communications and effective conflict resolution is thus an important component for achieving the success of the TCC and GMP procurement methods (Chan et al., 2002).

'Well-defined scope of work in client's project brief' (ranked fourth) and 'early involvement of the contractor in design development' (ranked sixth) are worthy of note as they were favourably scored as determinants for TCC and GMP success. Uncertainty regarding scope of work at the design development stage has been identified as the main inherent shortcoming of the TCC and GMP approaches.

Since there are frequent disputes between employers and contractors over whether employers' variations are regarded as design development or scope variation, the client's project brief should define clearly the scope of the contractor's work (Tang, 2005). As the design development may be continuously evolving in TCC and GMP procurement forms, interpretations regarding whether the changes are identified as TCC and GMP variations or arise out of design development could cause disputes if not promptly resolved (Gander and Hemsley, 1997). Hence, at the initial stage of a TCC or GMP construction project, the scope of work should be defined in as much detail and as accurately as possible, to minimise subsequent scope changes or necessary variations.

Furthermore, Sadler (2004) suggested that involving the expertise of the contractor at the early design stage brings benefits to target cost-type contracts. If the contractor is proactive and participates more at the design development stage, it may achieve better advanced works and programme planning, especially in materials procurement, and enhance the buildability of the project design remarkably. Their early involvement and influences on the pre-construction stage, construction methods and selection of materials

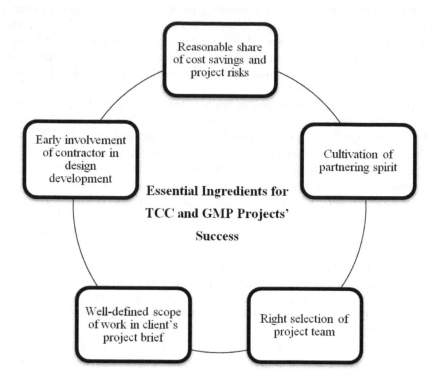

Figure 5.2 Essential ingredients for TCC and GMP projects' success (Chan et al., 2010c).

are essential to TCC and GMP success in optimising time, cost and quality (Hong Kong Housing Authority, 2006). This early involvement allows both employer and contractor to decide the suitable ownership of risks and encourages the achievement of a fair allocation of risks related to a TCC and GMP arrangement (Sadler, 2004). Valuable opinions on the key components driving the success of TCC and GMP projects were received based on the questionnaire survey, and are summarised in Figure 5.2.

Consultants concurred less on whether early involvement of the contractor at the design stage was an important contributor to TCC and GMP success, while contractors agreed more that their combination of expertise and innovative ideas for both design and construction could enhance the buildability of projects (Masterman, 2002; Hong Kong Housing Authority, 2006). These conspicuously differing views may be attributed to the traditional industry practice that the design is completed by an independent design consultant team because of its inherent competence and expertise, except for design-and-build contracts. As TCC and GMP contracts may diminish the role and significance of the team of design consultants, the consultant group may agree less on this factor than the contractors.

The selection of sharing ratios in TCC

Badenfelt (2008) identified the following major factors when negotiating the sharing ratio, based on the desktop literature review:

- the employer's and contractors' perceived degree of risk;
- their attitudes towards risk – contractors wish to have a high sharing proportion with a high uncertainty level, and vice versa;
- the importance of the contractor's incentives – the contractor's motivation to minimise cost depends on the sharing ratio;
- the initial understanding of the accuracy of the negotiated target cost among the contracting parties;
- the amount of the target profit – a larger profit encourages the builder to buy in;
- the desire of the contractor to increase the opportunities for profit in the longer run.

Badenfelt (2008) also conducted semi-structured interviews with clients and contractors regarding the sharing ratios.

Employer views on sharing ratios

Some project consultants believed that the contractors have to bear a larger sharing ratio in order to influence their motivation. Nevertheless, clients may respond that a large share for the contractor might lead to a harmful influence on project quality, as the builder/contractor may over-emphasise cost reductions and neglect the project goals or quality standards. They may evaluate different proposed sharing percentages which may calculate a sharing ratio, or invite the builders to suggest a sharing ratio, since the employer adopts competitive tendering. One client did not promulgate explicit rules when setting the sharing ratio, mentioning that a fixed sharing ratio or floating rates could be chosen. In respect of floating rates, the sharing rate will vary along with the project cost.

One employer focused on the beneficial aspects of TCC and an open-book accounting regime. However, he mentioned that the amount of share does not have an important influence on contractor incentives, opining that openness is essential to show all the cost data and the overall picture in order to make the appropriate decisions. The fraction of the share would probably be less significant. Three clients emphasised the link between the target cost and sharing ratio. One client mentioned that if 70% of the savings or 30% of the losses are allocated to the contractor, the builder will suggest an accurate target cost. This may provide the builder with a stronger motivation to propose a practical or genuine target cost. The client selected a project and calculated a mean value of four of the best tenders. The closest tender price to the mean value obtained the highest score. This arrangement prevents the tenderers from providing an unrealistic price during tender.

Another employer believed that as the builder has advantages in possessing considerable data in estimating target costs, a limit should be set to control how much the builder can gain. None of the employers considered any sophisticated way of forecasting the possible results of a target cost design. One employer mentioned relying on common sense. Nonetheless, two consultants suggested that they evaluated different sharing ratios, target costs and target fees in order to identify the optimal tender. One of them collected target cost and sharing ratios under various scenarios in order to calculate the most advantageous target cost, which might be a low target cost and a high sharing ratio, or vice versa. Nevertheless, their decisions were made based on a simple judgement process, without actual data to support the outcomes. One consultant utilised a computer program to stimulate various outcomes and predict the situation if the target cost was higher than the pre-agreed one. By adopting this relatively more sophisticated model, he could also set the ceiling prices and cut-off points.

He stated that the program was efficient and simple for working with an inexperienced employer who might only be concerned about the target cost, as it enables the employer to comprehend why selecting a builder who lodged a higher target cost could be financially beneficial.

Contractor views on sharing ratios

The sharing ratio was determined at the initial negotiation stage (mentioned by two contractors). Generally, sharing ratios ranged from 0.3 to 0.7. One builder mentioned that a 0.5 sharing ratio was equitable, but the other contractors did not place much emphasis on fairness.

One contractor opined that a sharing ratio of 0.5 was the most efficient. When working for the client, a larger proportion of the risk is moved to the builder, so that the contractor bears 70% of the delay. However, the contractor would be encouraged to increase the target cost, as it would include the risk premium in the tender. One contractor stated that the more detailed the design at the tender stage, the more discussions of target cost variations there could be when executing the project. These discussions are less prevalent when the builder is involved in the early project design phase.

Chapter summary

Chan et al. (2007a) identified the following critical success factors for TCC and GMP projects:

- a standard form of contract for TCC and GMP projects;
- well-defined scope of work stated in the client's project brief;
- knowledge and experience of TCC and GMP methodology among the client, consultants, main contractor and subcontractors;
- an appropriate selection of the project team;

- equitable sharing of cost saving and fair allocations of risks;
- a partnering spirit among all contracting parties;
- early involvement of the contractor in developing the design;
- formation of adjudication committee, and meeting to settle any issues and disputes;
- a proactive main contractor throughout the TCC and GMP project, to manage any intractable problems;
- an open-book accounting regime offered by the main contractor to support its tender pricing.

Part IIA

Practices of target cost contracting

6 Key risk factors in implementing target cost contracts

Introduction

Risk is defined as 'the chance of something happening that will have an impact upon the objectives' (AS/NZS3460:2004 – Risk Management). A considerable amount of literature shows the essential components of risk management. Risk is related to the probability and results of a task, and the overall effect on project objectives (Environment, Transport and Works Bureau, 2005). Risk can be controlled, reduced, accepted or shifted. However, it cannot be neglected in project delivery (Latham, 1994). The purposes of risk management are to ensure that: (1) risks are transferred to the stakeholder who can best manage it; (2) risks are allocated impartially and reasonably; and (3) there is allowance for unpreventable cost due to the risks which are perceived to exist somewhere throughout the project life (Ahmed et al., 1998). Risk management, involving risk planning, risk identification, risk analysis, risk evaluation and risk treatment, is supported by continuous supervision, reviewing and documenting of the identified risks, as well as effectual communication and consultation among contracting teams (Environment, Transport and Works Bureau, 2005). Risk management procedures are illustrated in Figure 6.1.

As the HKSAR Government is a major client of the Hong Kong construction industry, a systematic risk management process set out by the Environment, Transport and Works Bureau (2005) is implemented in all works departments under the Development Bureau (the Environment, Transport and Works Bureau was renamed the Development Bureau in 2008), while risk management in the private sector is less structured and is mostly based on experience and intuition. Thus, this systematic risk management regime implemented by the HKSAR Government was selected as the primary topic for this chapter. As there are various common risks in construction projects generally, such as Act of God, this chapter first identifies the major risk factors under TCC and GMP procurement methods.

Establishment of context and risk planning

The major component of the risk management procedure is setting up the context and risk planning. The evaluation criteria and the structure

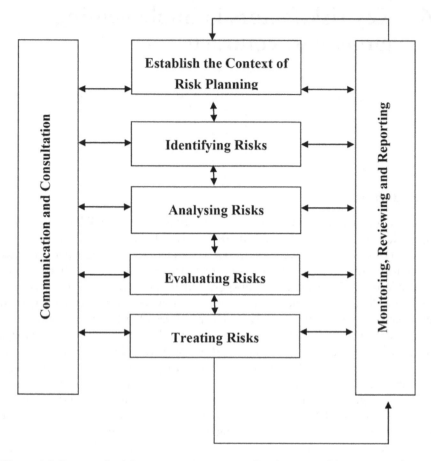

Figure 6.1 Systematic risk management process (Environment, Transport and Works Bureau, 2005).

of analysis to identify the risks should be defined and well established (Environment, Transport and Works Bureau, 2005). Risk management needs to be carried out based on the intent and the structure of risk management procedure as identified in the risk management plan.

Risk identification

The aim of risk identification is to figure out what the sources and types of risk impacting the achievement of project objectives are (Flanagan and Norman, 1993; Environment, Transport and Works Bureau, 2005). The classifications of risks includes manageable and unmanageable risks, dependent and independent risks, project and individual risks, etc. (Flanagan and Norman, 1993). There are several approaches that could be used for risk identification, solely or in conjunction, such as brainstorming in a workshop environment, using a risk register checklist or consulting specialists in a particular field (Environment, Transport and Works Bureau, 2005). The adoption of these approaches depends on the nature of project as well as the resources available.

Risk analysis

Subsequent to risk identification, it is necessary to analyse and quantify the risks. The purpose of risk analysis is to develop a perception of the degree of risk and its nature for managers. It can help decision makers to consider what risks need to be managed and how to do so. The level of risk is estimated based on two measurements: likelihood of occurrence and the impact of the risk. A risk analysis matrix can be an effective method for risk analysis (Environment, Transport and Works Bureau, 2005).

Risk analysis can be carried out with sufficient details of the risks, the particulars, data and resources available. Risk analysis can be classified as qualitative analysis and quantitative analysis, and both of these approaches in broad terms (Environment, Transport and Works Bureau, 2005). Nevertheless, the restrictions of using qualitative analysis, involving expert judgement, are emphasised by several scholars who have highlighted the importance of using formal methods to assess risks and make appropriate decisions (Flanagan and Norman, 1993; Chege and Rwelamila, 2000). Quantitative analysis can also be used to comprehend further benefits, such as clear contingency setting, controlling draw-down of contingency and insurance. Some quantitative techniques for risk analysis are based on anticipated monetary value, decision tree diagrams, sensitivity analysis and simulation techniques.

Risk evaluation

The objective of risk evaluation is to acquire an understanding of the degree of risk and provide some information to project managers to decide the actions required to manage the risks and identify priorities for these actions. Table 6.1 shows a set of risk evaluation criteria suggested by the Environment, Transport and Works Bureau (2005) to demonstrate how the decisions could be made.

Risk treatment

As indicated in Flanagan and Norman (1993), risk treatment can be divided into the following forms:

1 risk retention;
2 risk reduction;
3 risk transfer (risk sharing);
4 risk avoidance.

Table 6.1 Sample set of risk evaluation criteria (Environment, Transport and Works Bureau, 2005)

Level of risk	Recommended level of management attention
Extreme	Immediate actions from top management needed, development of action plans required with clear assignment of responsibility and timeframe for each party
Very high High	Top management attention required, development of action plans required with clear assignment of responsibility and timeframe for each party
Medium	Risk requires specific ongoing monitoring and review, to make sure that the level of risk does not rise further; otherwise, managed by routine procedures
Low	Risk can be accepted or even ignored; managed by routine procedures, unlikely to need specific application of resources

Selection of the above four risk treatment methods is de facto a cost-benefit decision, with preference given to treatments providing the best outcome for the project (Environment, Transport and Works Bureau, 2005).

Risk monitoring and review

Since construction technology keeps advancing, management has become increasingly complicated with respect to projects and organisational structure over the decades, leading to a dynamic risk management process. The extent of impact and the possibility of occurrence of the risk may vary throughout the project delivery process (Environment, Transport and Works Bureau, 2005). New risks may also emerge throughout the project. It is essential to ensure that the risk management process is alive and responsive to progress over the course of the project. Since the risks change over time, it is necessary to monitor them and the efficiency of risk treatments. The Environment, Transport and Works Bureau (2005) further suggested conducting regular risk reviews at major milestone stages during the project delivery process.

Risk communication and consulting

As shown in Figure 6.1, communication and consulting are included in every procedure of the risk management process. Internal project teams can contribute to risk management by direct involvement in workshops or information distribution. It is highly recommended that they communicate with each other and contribute to the process of risk management in order to achieve continuous improvement throughout the construction project (Environment, Transport and Works Bureau, 2005).

Classification of risk factors in TCC and GMP projects

Although there are a number of perceived benefits of applying the TCC and GMP approaches to projects, the improper use of TCC and GMP procurement can generate lots of problems and bring considerable risks to both contractual parties.

Ashley et al. (1989) pointed out that economic pressure could lead to adversarial relationships between the employer and contractor. This could limit the contractor's input on design constructability and incentives to improve quality. Hence, clients consider that economic pressure may be one of risk factors for GMP construction projects.

As suggested by Al-Harbi and Kamal (1998), with various attitudes towards risks, the client and builder value the sharing proportion and final project cost differently. A sharing fraction that best suits both parties is not easy to achieve. Imbalance in the sharing portion, which would impact incentives for the contractor, may be considered as one of the risk factors under this procurement method.

Yew (2008) mentioned that an unclear tender design brief, vaguely identified contractor design accountabilities, ambiguous liability for errors and omissions, as well as insufficient builder involvement during design development will raise the risk exposure of builders under GMP contracts. Indeed, many risk factors in typical construction projects are documented in the literature, as shown in Table 6.2.

Major risk factors of TCC

Chan (2011) conducted an empirical questionnaire survey to examine the major risk factors when implementing TCC. Table 6.3 shows the descending ranking of the perceived risks and highlights the top ten most significant risks in bold. The most important risk indicated in Table 6.3 was 'change in scope of work'. This finding echoes Septelka and Goldblatt (2005), who studied change order data of forty-six construction projects under GMP contracts in the United States. They found that change in scope of work accounted for about half of the cost change in construction projects. Cox et al. (1999) investigated some cases in the United Kingdom and found that design changes could be ascribed to changes in clients' requirements in the cases they explored.

The second most significant risk was 'insufficient design completion during tender invitation'. In the building industry in Hong Kong, the schedule for project development is very tight, so the design is still immature at tender invitation stage. It is unavoidable that variation orders are issued by the architect or engineer after contract award. It is difficult to determine whether a variation should be considered as a change in scope of work, or as a design development which does not change the target cost or GMP value (Haley and Shaw, 2002). Yew (2008) shared a similar perception that this may lead to disputes during the construction phase (i.e. post-contract stage)

Table 6.2 Risk factors associated with typical construction projects (Chan et al., 2008)

Risk factors	Bernhard 1988	Ahmed et al. 1998	Al-Harbi and Kamal 1998	Ahmed et al. 1999	Broome and Perry 2002	Haley and Shaw 2002	Rahman and Kumaraswamy 2002	Cheng 2004	Fan and Greenwood 2004	Oztas and Okmen 2004	Sadler 2004	ETWB 2005	Li et al. 2005	Tang 2005	HKHA 2006	Shen et al. 2006	Ng and Loosemore 2007	Chan et al. 2007a	Chan et al. 2007b	Yew 2008	Total no. of hits for a certain risk factor
Act of God				✓			✓					✓	✓			✓	✓				6
Adequacy of design		✓										✓									2
Buildability/constructability							✓					✓									2
Change in government regulations				✓			✓					✓	✓								4
Client may pay more as contractor would inflate the tender sum to cover additional risks								✓						✓	✓			✓	✓		5
Conflict of documents				✓																	1
Contractor may incur a loss due to unclear scope of work						✓		✓	✓					✓							4
Contractor may not foresee design development risks							✓	✓	✓					✓			✓	✓	✓	✓	8
Defective design				✓			✓														2
Delayed payment on contracts				✓			✓														2
Delay in availability of labour, materials and equipment				✓			✓			✓			✓					✓			5
Delay in resolving contractual issues							✓			✓											2
Design changes											✓							✓			2
Difficult to value revised contract price												✓	✓					✓		✓	4

Risk	No.
Difficult to use successfully on contracts where many changes are expected	1
Disputes may arise due to change in scope of work	9
Exchange rate variation	1
Errors and omissions in tender documents / Insufficiency of tender	6
Financial failure of contractor	3
Financial failure of owner	2
GMP may not be the 'maximum' at end of the day	2
Inaccurate topographical data	1
Inclement weather	4
Inexperienced contractor may jeopardise the GMP/TCC process	7
Inflation	6
Labour and equipment productivity	2
No standard form of contract leads to misunderstanding of responsibilities of parties	5
Oil/energy/commodity price fluctuation	1
Quality of work	2
Subcontractor failure	3
Third-party delay (risk)	4
Uneven sharing fraction of saving/overrun of budget	3
Unforeseen ground conditions	7
Total no. of risks identified from each publication	1 5 1 15 1 5 22 4 3 7 2 13 6 6 3 2 5 9 3 5 **118**

Note: The previous studies are ranked in increasing chronological order of year of publication, followed by alphabetical order of surnames of authors. ETWB = Environment, Transport and Works Bureau; HKHA = Hong Kong Housing Authority.

Table 6.3 Impacts of risk factors encountered with TCC and GMP schemes by all respondents (Chan, 2011)

No.	Risk factor	Mean	Standard deviation	Rank
5	Change in scope of work	16.41	8.26	1
17	Insufficient design completion during tender invitation	15.46	7.38	2
20	Unforeseeable design development risks at tender stage	14.54	7.20	3
6	Errors and omissions in tender document	14.51	7.52	4
21	Exchange rate variations	14.49	7.39	5
29	Unforeseeable ground conditions	14.25	7.68	6
1	Actual quantities of work required far exceeding estimate	13.97	8.09	7
32	Lack of experience of contracting parties throughout TCC and GMP process	13.91	7.74	8
22	Inflation beyond expectation	13.81	7.04	9
3	Unrealistic maximum price or target cost agreed in the contract	13.76	7.95	10
4	Disagreement over evaluating the revised contract price after submitting an alternative design by main contractor	13.51	7.42	11
7	Difficult for main contractor to have back-to-back TCC and GMP contract terms with nominated or domestic subcontractors	13.31	8.53	12
26	Global financial crisis	13.19	8.19	13
18	Poor buildability/constructability of project design	13.11	6.56	14
2	Delay in resolving contractual disputes	13.11	7.21	15
9	Loss incurred by main contractor due to unclear scope of work	13.07	7.24	16
16	Delay in work due to third party	12.64	5.89	17
28	Inclement weather	12.43	7.37	18
8	Inaccurate topographical data at tender stage	12.40	7.25	19
19	Little involvement of main contractor in design development process	12.36	7.66	20
15	Selection of subcontractors with unsatisfactory performance	12.17	6.43	21
31	Difficult to obtain statutory approval for alternative cost-saving designs	12.16	6.43	22
33	Impact of construction project on surrounding environment	12.15	7.43	23
12	Poor quality of work	12.07	7.53	24
11	Technical complexity and design innovations requiring new construction methods and materials from main contractor	11.92	7.22	25
23	Market risk due to the mismatch of prevailing demand of real estate	11.86	6.98	26
24	Change in interest rate on main contractor's working capital	11.33	6.87	27
34	Environmental hazards of constructed facilities towards the community	11.17	6.97	28

13	Delay in availability of labour, materials and equipment	11.03	6.10	29
25	Delayed payment on contracts	10.81	6.82	30
30	Change in relevant government regulations	10.80	6.48	31
10	Difficult to agree on a sharing fraction of saving/overrun of budget at pre-contract award stage	10.72	6.57	32
14	Low productivity of labour and equipment	10.09	5.68	33
27	*Force majeure* (Acts of God)	8.66	6.74	34

concerning whether the developing or developed project design and its refinement accounting for an improvement of the initial design intent or a change in client's requirements constitute a variation and change GMP value.

'Unforeseeable design development risks at tender stage' was discerned as the third most significant risk factor under TCC and GMP procurement. The main contractor needs to abide by the contract sum when further developing the incomplete design provided during tendering. In other words, contractors are required to complete undefined works. If the quantities are underestimated by contractors at the design development stage, they will probably suffer a monetary loss. Yew (2008) commented that builders had to bear all the risks involved in GMP agreements, such as defects of the original design in the tender. Davis Langdon and Seah (2003) mentioned that it is risky for both client and contractor if the agreement on GMP is achieved too early with insufficient design details. Fan and Greenwood (2004) observed that the design development stage under GMP approaches could lead to contractual disputes. Oztas and Okmen (2004) suggested that employers should establish a list of comprehensive employers' requirements and state them in the tender documents in order to prevent excessive and unnecessary changes in design at the post-contract stage. This risk may be attributable to an inadequate tendering period, which means that contractors cannot fully comprehend the scope of work and identify the potential pitfalls involved in the contract.

The fourth most notable risk under TCC and GMP procurement strategies was 'errors and omissions in tender document'. The contract tender documents are the major elements for risk allocation. A considerable number of intractable conflicts and disputes as well as dispensable contract variations after contract award may result if there are omissions, errors or discrepancies within the contract document at project commencement. Laryea (2011) noted that ambiguity in tender documents can be a major source of claims and disputes during the post-contract stage. Yew (2008) agreed with the finding that builders were required to bear all of the risks associated with TCC and GMP schemes, involving omissions and errors in tender documents.

The fifth most important risk under TCC and GMP contracts was 'exchange rate variations'. However, Tam et al. (2007) stated that 'exchange rate variations' was not regarded as a major risk in Hong Kong. Apparently, the high significance of this risk factor may be attributable to the time of the study, near the financial crisis, so industry practitioners were more concerned with exchange rates.

Chapter summary

This chapter has discussed risk management and identified the major risk factors and their corresponding significance in risk management of TCC and GMP projects. It is important to understand the significant risks in the delivery of TCC and GMP projects, especially change in scope of work, incomplete design when inviting tenders, and unexpected design developments. If these risk factors are managed properly, value for money will be enhanced in the whole procurement stage.

7 Risk assessment model for target cost contracts

Introduction

Scholars have published a great number of academic papers on TCC and GMP which align the objectives of clients and contractors. Nonetheless, there has been limited research focusing on risk identification and assessment of these kinds of projects. In this chapter, the development of a risk assessment model for TCC and GMP arrangements will be described using factor analysis and the fuzzy synthetic evaluation method, based on the previous questionnaire survey (Chan, 2011), in order to arrive at a more objective risk assessment. An Overall Risk Index (ORI) of a project and risk indices of individual principal risk groups (PRGs) can be established using the model. The development of this model helps the project team understand how to achieve a successful TCC and GMP projects. This also provides a platform for evaluating the risk level of projects on the basis of objective proof instead of intuitive evaluations. The results shown in this chapter have been published in a journal article (D.W.M. Chan et al., 2011a).

Risk assessment in construction projects

Taroun (2014) carried out a comprehensive literature review on risk assessment methods and modelling in the construction sector. Indeed, construction is a risky business because of the one-off nature of construction projects and the involvement of various stakeholders (Flanagan and Norman, 1993). Laryea and Hughes (2008) interviewed five estimators in United Kingdom construction companies and found that a number of them utilised a risk register mechanism, comprising risk impact as a function of probability multiplied by severity (i.e. a probability–severity model). The risk assessment commenced with a brainstorming review workshop in which the respondents defined the project risks. Subsequently, the risks were assessed using a spreadsheet matrix to help calculate the contingencies according to the severity value and the probability value based on the hands-on experience and intuitive judgement of the respondents. Adams (2008) opined that risks in construction projects were usually evaluated using an arbitrary approach.

Builders often add a single arbitrary cost contingency to show their overall perception of the total risks, but do not evaluate the risks they are required to bear. This finding was in line with a previous study by Akintoye and Macleod (1997), who studied how builders analysed risk using a questionnaire survey in the United Kingdom. Akintoye and Macleod (1997) and Adams (2008) drew a similar conclusion that formal risk analysis techniques were seldom applied in the industry due to doubts about suitability and insufficient experience of those techniques.

Risk management brings advantages to project development when management has applied a systematic approach from planning to completion in order to help project stakeholders to make more efficient and informed decisions (Baloi and Price, 2003). The unstructured and arbitrary nature of risk management can harm projects. Truly, risk management is both a science and an art (Baloi and Price, 2003). Although there are a considerable number of research studies and continuous effort devoted to risk management in terms of construction management, industrial practitioners seem to neglect their importance (Flanagan and Norman, 1993). Unlike other industries, such as the petrochemical and oil industries, there is still a remarkable gap between current practices and existing theories in risk management in the construction industry (Thompson and Perry, 1992).

Recently, many risk assessment models have been established to enrich knowledge of construction risk management. For instance, Baloi and Price (2003) established a fuzzy decision framework to model the global risk factors affecting construction cost. Zhang and Zou (2007) generated a risk assessment model for joint venture projects in mainland China using the Fuzzy Analytical Hierarchy Process. Ng et al. (2007) established a simulation model enabling a public partner (government) to determine the concession period of a public–private partnership scheme by considering the expected investment and tariff. Zeng et al. (2007) utilised fuzzy reasoning techniques to establish a tool for managing risks in construction projects. Nevertheless, there have been few if any risk assessment models for TCC and GMP strategies. These results from a desktop search have further reinforced the fundamental objective of this study: to establish a risk assessment model for TCC and GMP projects in Hong Kong.

Development of a fuzzy risk assessment model (FRAM)

Overall research framework

Literature review

Figure 7.1 shows the overall research framework for establishing the fuzzy risk assessment model (Chan, 2011). Risk management is an essential component of TCC and GMP projects. The first step in risk management is risk identification. There are thirty-four individual key risk factors of TCC and

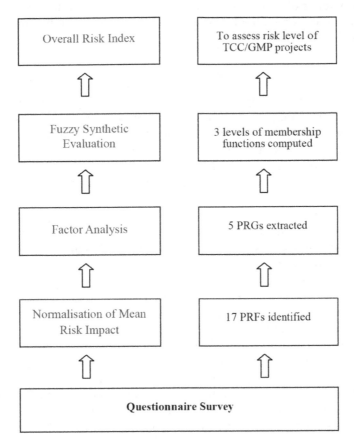

Figure 7.1 Overall research framework adopted in developing the fuzzy risk assessment model (Chan, 2011).

GMP projects, according to an exhaustive literature review, including the relevant textbooks, journal articles, research reports, conference proceedings, corporate newsletters, dissertations and online resources. Hence, the list of thirty-four key risk factors relating to TCC and GMP strategies was regarded to be relevant, sufficient and representative.

Factor analysis

The objective of factor analysis is to reduce a large number of observed variables to a more manageable number of factors with a minimum loss of information, and to show the inter-relationships among variables (Hair et al., 1998). Principal components analysis was conducted for factor analysis, and the Equamax rotation method with Kaiser normalisation was performed. The objective of principal components analysis was to derive a smaller number of variables in order to convey as much data on the seventeen risk factors crystallised by normalisation of the combined mean scores as possible.

Fuzzy synthetic evaluation method

It has been generally accepted that risk assessment is a typical multi-objective problem, as it is influenced by many uncertainties and variations. To facilitate the decision making process for the uncertainties and variations, fuzzy set theory and the fuzzy synthetic evaluation model were applied to generate the risk assessment model in the study (Zhang et al., 2004).

A fuzzy synthetic evaluation model needs three fundamental elements:

1   a set of basic criteria/factors $\pi = \{f_1, f_2 \ldots \ldots f_{17}\}$, e.g. f_1 = actual quantities of work required far exceeding estimate; f_2 = delay in resolving contractual disputes; $\ldots \ldots f_{17}$ = insufficient design completion during tender invitation;
2   a set of grade alternatives $E = \{e_1, e_2, \ldots \ldots e_n$, e.g. e_1 = very low, e_2 = low, e_3 = moderate, e_4 = high and e_5 = very high (for severity); and e_1 = very very low, e_2 = very low, e_3 = low, e_4 = moderate, e_5 = high, e_6 = very high and e_7 = very very high (for likelihood);
3   for every object $u \notin U$ (which means the fuzzy subset u does not belong to the fuzzy set U), there is an evaluation matrix $R = (r_{ij})_{m \times n}$. Under the fuzzy environment, r_{ij} is the degree to which alternative e_j satisfies the criterion f_i. It is presented by the fuzzy membership function of grade alternative e_j with respect to the criterion f_i. With the preceding three elements, for a given $u \notin U$, its evaluation result can be derived.

Risk assessment of TCC and GMP projects includes many PRFs and PRGs. All PRFs and PRGs have to be considered in order to enhance the risk assessment. Hence, it would be desirable if the synthetic evaluation method in this study could solve the problems with multi-attributes and multi-levels. Fuzzy synthetic evaluation, as one of the applications of fuzzy set theory, was used in this research to establish a fuzzy risk assessment model for TCC and GMP projects in Hong Kong.

Tah and Carr (2000) found that fuzzy sets can be applied to quantify linguistic variables and severity for risk assessment of construction projects. Zhang et al. (2004) revealed that it is always problematic to define uncertain information input for construction-oriented discrete-event simulation. Therefore, they decided to incorporate fuzzy set theory with discrete-event simulation to manage the vague, imprecise and subjective estimation of activity durations, especially when insufficient or even no sample data are available. In addition, Baloi and Price (2003) employed fuzzy set theory in their research and developed a decision framework for builders to manage global risk factors influencing cost performance.

Knight and Fayek (2002) suggested that fuzzy set theory can be applied to model construction issues where the process has only been available in the minds of experienced practitioners (Knight and Fayek, 2002). Both Chen and Cheng (2009) and Zeng et al. (2007) concurred that fuzzy set theory should be adopted when managing unclear information due to its ability to utilise natural language in terms of linguistic variables. As risk assessment for the construction industry mainly depends on individual intuition and experience (Flanagan and Norman, 1993), the method of fuzzy synthetic evaluation can be used in this study to manage the use of linguistic variables (e.g. 'high' and 'very high') and to allow for ranking or subjective rating of different risk factors (Knight and Fayek, 2002; Singh et al., 2008).

This approach was also employed in other areas. For instance, Lu et al. (1999) used fuzzy synthetic evaluation when analysing water quality in Taiwan, and expressed the change in water quality in their evaluation. Singh et al. (2008) adopted the same approach for assessing whether groundwater would be potable in India. As subjective judgements of evaluators are included when assessing the risks of TCC and GMP contracts, fuzzy synthetic evaluation is regarded as an appropriate tool to establish the risk assessment model for TCC and GMP schemes because the risk evaluation is usually multi-layered (Sadiq et al., 2005), fuzzy in nature and includes subjective judgements.

Selection of principal risk factors by normalisation of combined mean scores

As is widely accepted, the influence of a risk can be calculated by multiplying its level of severity and probability of occurrence (Cox and Townsend, 1998; Garlick, 2007). The impact of the thirty-four risks identified in the study was input in this way. Table 7.1 shows the combined scores for the thirty-four risks. The normalised value of each risk factor is derived by the following formula:

normalised value of impact = (average actual value – average minimum value)/(average maximum value – average minimum value)

Muller and Turner (2010) analysed the leadership competency profiles of successful project managers in different types of projects using a questionnaire survey. The mean of normalised scores of competencies served as a demarcation point to group the profiles of project managers in their survey. This study adopted the same logic, and the mean of all of the thirty-four normalised values was 0.49. Therefore, the value of 0.49 acts as the demarcation point for normalisation to choose only those upper half 'more important' risk factors with normalised values equal to or larger than 0.49 for conducting the subsequent factor analysis. Table 7.1 shows the risk

Table 7.1 Result of normalisation of risk factors by mean combined scores (Chan, 2011)

No.	Risk factor	Impact = severity × likelihood			Rank	Normalised value
		Severity	Likelihood	Impact		
5	Change in scope of work	3.53	4.48	15.84	1	1.00
17	Insufficient design completion during tender invitation	3.47	4.30	14.93	2	0.88
20	Unforeseeable design development risks at tender stage	3.38	4.13	13.98	3	0.74
6	Errors and omissions in tender document	3.44	4.05	13.97	4	0.74
21	Exchange rate variations	3.31	4.19	13.86	5	0.73
29	Unforeseeable ground conditions	3.50	3.93	13.76	6	0.71
3	Unrealistic maximum price or target cost agreed in the contract	3.66	3.64	13.30	7	0.65
1	Actual quantities of work required far exceeding estimate	3.46	3.83	13.27	8	0.65
22	Inflation beyond expectation	3.34	3.91	13.08	9	0.62
32	Lack of experience of contracting parties throughout TCC and GMP process	3.30	3.93	12.99	10	0.61
26	Global financial crisis	3.70	3.50	12.94	11	0.60
4	Disagreement over evaluating the revised contract price after submitting an alternative design by main contractor	3.21	4.02	12.93	12	0.60
18	Poor buildability/constructability of project design	3.40	3.77	12.82	13	0.59
2	Delay in resolving contractual disputes	3.28	3.88	12.72	14	0.57
9	Loss incurred by main contractor due to unclear scope of work	3.46	3.62	12.54	15	0.55
7	Difficult for main contractor to have back-to-back TCC and GMP contract terms with nominated or domestic subcontractors	2.97	4.21	12.49	16	0.54
16	Delay in work due to third party	3.24	3.81	12.37	17	0.52
28	Inclement weather	2.92	4.11	12.01	18	0.47
8	Inaccurate topographical data at tender stage	3.24	3.65	11.84	19	0.45

15	Selection of subcontractors with unsatisfactory performance	3.34	3.52	11.76	20	0.44
19	Little involvement of main contractor in design development process	2.98	3.92	11.68	21	0.43
31	Difficult to obtain statutory approval for alternative cost-saving designs	3.16	3.69	11.65	22	0.42
33	Impact of construction project on surrounding environment	3.11	3.69	11.48	23	0.40
11	Technical complexity and design innovations requiring new construction methods and materials from main contractor	3.18	3.57	11.35	24	0.38
12	Poor quality of work	3.19	3.53	11.25	25	0.37
23	Market risk due to the mismatch of prevailing demand of real estate	3.06	3.64	11.14	26	0.35
24	Change in interest rate on main contractor's working capital	2.97	3.54	10.50	27	0.27
13	Delay in availability of labour, materials and equipment	3.10	3.37	10.46	28	0.26
34	Environmental hazards of constructed facilities towards the community	3.04	3.40	10.34	29	0.24
10	Difficult to agree on a sharing fraction of saving/overrun of budget at pre-contract award stage	3.06	3.37	10.29	30	0.24
30	Change in relevant government regulations	3.00	3.42	10.27	31	0.23
25	Delayed payment on contracts	3.07	3.31	10.14	32	0.22
14	Low productivity of labour and equipment	3.02	3.19	9.63	33	0.15
27	Force majeure (Acts of God)	3.24	2.64	8.56	34	0.00

factors and their corresponding normalised values. A total of seventeen risk factors with normalised values equal to or exceeding 0.49 (shown in bold) were selected for conducting factor analysis.

The major reason for performing such normalisation of mean scores is that it can extract a more manageable number of risk factors, because industrial practitioners tend to use a more user-friendly model with fewer inputs for application (Yeung et al., 2007). According to Gorsuch (1983) and Lingard and Rowlinson (2006), such selection also conforms to the prerequisite of the factor analysis technique, which requires a ratio of 1:5 for variables under study to sample size.

In addition, normalisation helps to develop the model in terms of practicality and user-friendliness. It is recommended to lower the number of risk factors, as users are required to input their perceived severity and likelihood of each risk factor involved in the model to develop the Overall Risk Index and identify the most critical principal risk group. If all the risk factors are involved, the users are required to input 68 items (34 risk factors in total × 2 entities for each risk factor) in the model to acquire the results. It would be inconvenient for industrial practitioners to enter such a huge amount of data in the risk assessment model. It is therefore essential to perform normalisation to strike a balance between the comprehensiveness of risk factors involved in the model and practicality/user-friendliness of the model.

Table 7.1 shows that the eliminated risk factors are not enough within the construction industry. For instance, Item 25, 'delayed payment on contracts', is omitted after normalisation because builders are protected from delayed payment by the clients by standard forms of contracts in both building and civil engineering works in Hong Kong (under Clause 79(4)(a) of the General Conditions of Contract for Building Works of the HKSAR Government and Clause 51.2 of NEC3 Option C: Target Cost with Activity Schedule). The respondents may consider that non-payment may influence all types of contracts given to a defaulting employer, so Item 25 was not regarded as a severe risk, supporting by El-Sayegh (2008) and Ahmed et al. (1999).

Item 13, 'delay in availability of labour, materials and equipment' was also eliminated after normalisation. This finding is in line with the study of Tam et al. (2007), which shows that the same factor was ranked seventeenth out of twenty-four risk factors by the Architect respondent group and four-teenth by the Engineer respondent group in their questionnaire survey on risks in public foundation projects in Hong Kong.

Furthermore, Item 30, 'change in relevant government regulations', was eliminated after the normalisation exercise. This finding matches the survey result of Tam et al. (2007) that 'change in statutory requirements' was ranked twentieth out of twenty-four risk factors by both the Architect group and Engineer group. Therefore, the survey results of this study echo those of previous studies and are considered logical and reasonable.

Identification of five principal risk factors for TCC and GMP projects in Hong Kong

The influence of a single risk factor was assessed based on the multiple of the probability of occurrence and the level of severity. According to the results of the normalisation, taxonomy was established with factor analysis which analysed the structure of inter-relationship among data by identifying a set of common underlying concepts known as factors (Rosch, 1988). Chan (2011) found that factor analysis can be performed with confidence in this study.

The Equamax rotation method with Kaiser normalisation was performed using the SPSS FACTOR program. The method of Equamax rotation generates the highest individual factor loadings for the same set of individual factors, and more illustrative overall results as used and suggested by both Abraham et al. (1994) and Emsley et al. (2003). The objective of principal components analysis is to derive a smaller number of variables in order to convey as much information about the top seventeen key risk factors crystallised by normalisation of combined mean scores as possible out of a total of thirty-four.

Table 7.2 shows the loadings of each factor to demonstrate the relationships between the risks and the PRGs. These loadings indicate to what extent the risks influence the establishment of PRGs. Five PRGs were extracted, as shown in Table 7.2, accounting for 69% of the variance in responses. They consist of: (1) 'third-party delay and tender inadequacy', (2) 'post-contract risks', (3) 'lack of experience in TCC and GMP process', (4) 'design documentation risks' and (5) 'economic and financial risks'.

Development of weightings of the seventeen PRFs and five PRGs for TCC and GMP projects in Hong Kong

The next stage in establishing the fuzzy risk assessment model for TCC and GMP construction projects is to derive the appropriate weightings for each PRF and PRG. According to Yeung et al. (2007), the weightings for each of the seventeen PRFs and 5 PRGs can be derived using the following formula:

$$W_i = \frac{M_i}{\sum_{i=1}^{5} M_i}$$

where:

W_i represents the weighting of a particular PRF or PRG;
M_i represents the mean ratings of a particular PRF or PRG;
$\sum M_i$ represents the summation of mean ratings of all the PRFs or PRGs.

Table 7.3 shows the corresponding weightings for each of the seventeen PRFs and five PRGs.

Table 7.2 Principal risk groups of TCC and GMP projects extracted by factor (Chan, 2011)

No.	Item	Factor loading	Eigenvalue	% of variance explained	Cumulative % of variance explained
PRG 1 – Pre-contract risks					
16	Delay in work due to third party	0.670	6.857	40.338	40.338
4	Disagreement over evaluating the revised contract price after submitting an alternative design by main contractor	0.645			
3	Unrealistic maximum price or target cost agreed in the contract	0.577			
20	Unforeseeable design development risks at tender stage	0.575			
18	Poor buildability/constructability of project design	0.554			
17	Insufficient design completion during tender invitation	0.498			
PRG 2 – Post-contract risks					
7	Difficult for main contractor to have back-to-back TCC and GMP contract terms with nominated or domestic subcontractors	0.849	1.500	8.822	49.159
1	Actual quantities of work required far exceeding estimate	0.718			
2	Delay in resolving contractual disputes	0.634			
9	Loss incurred by main contractor due to unclear scope of work	0.501			
PRG 3 – Lack of experience in TCC and GMP procurement process					
32	Lack of experience of contracting parties throughout TCC and GMP process	0.878	1.217	7.157	56.316
PRG 4 – Design risks					
5	Change in scope of work	0.821	1.119	6.581	62.897
29	Unforeseeable ground conditions	0.557			
6	Errors and omissions in tender document	0.483			
PRG 5 – Economic and financial risks					
26	Global financial crisis	0.783	1.046	6.154	69.051
21	Exchange rate variations	0.727			
22	Inflation beyond expectation	0.606			

Table 7.3 Weightings for the seventeen principal risk factors and five principal risk groups for TCC and GMP construction projects in Hong Kong (Chan, 2011)

Risk factor (RF)	Risk level of severity				Risk likelihood of occurrence			
	Mean for severity	Weighting for each PRF	Total mean for each PRG	Weighting of each PRG	Mean for likelihood	Weighting for each PRF	Total mean for each PRG	Weighting of each PRG
RF 16	3.24	0.16			3.81	0.16		
RF 4	3.21	0.16			4.02	0.17		
RF 3	3.66	0.18			3.64	0.15		
RF 20	3.38	0.17			4.13	0.17		
RF 18	3.40	0.16			3.77	0.16		
RF 17	3.47	0.17			4.30	0.19		
PRG 1 – Pre-contract risks			20.36	0.35			23.67	0.35
RF 7	2.97	0.23			4.21	0.27		
RF 1	3.45	0.26			3.83	0.25		
RF 2	3.28	0.25			3.88	0.25		
RF 9	3.46	0.26			3.62	0.23		
PRG 2 – Post-contract risks			13.16	0.23			15.54	0.23
RF 32	3.30	1.00			3.93	1.00		
PRG 3 – Lack of experience in TCC and GMP procurement process			3.30	0.06			3.93	0.06
RF 5	3.53	0.34			4.47	0.36		
RF 29	3.50	0.33			3.93	0.32		
RF 6	3.44	0.33			4.05	0.33		
PRG 4 – Design risks			10.47	0.18			12.45	0.19
RF 26	3.70	0.36			3.50	0.30		
RF 21	3.31	0.32			4.19	0.36		
RF 22	3.34	0.32			3.91	0.34		
PRG 5 – Economic and financial risks			10.35	0.18			11.60	0.17
Total			57.64	1.00			67.19	1.00

Computation of membership function of each PRF and PRG

Seventeen PRFs were acquired from normalisation of the combined mean scores for gauging the overall risk level of TCC and GMP construction projects in Hong Kong. Assume that the set of basic criteria applied in the fuzzy risk assessment model is $\pi = \{f_1, f_2 \ldots \ldots, f_{17}\}$; and the grades for selection are identified as $E = \{1,2.3,4,5\}$, where 1 = very low, 2 = low, 3 = moderate, 4 = high and 5 = very high (for severity); and $E = \{1,2.3,4,5,6,7\}$, where 1 = very very low, 2 = very low, 3 = low, 4 = moderate, 5 = high, 6 = very high and 7 = very very high (for likelihood). In respect of each particular PRF, the membership function is formed by the evaluation of survey participants. For instance, the survey results on 'actual quantities of work required far exceeding estimate' showed that 2% of the respondents commented that the level of severity of this risk to the project was very low, 17% low; 33% moderate; 30% high and 18% very high, thus the membership function of this risk is developed as:

$$C1 = \frac{0.02}{\text{very low}} + \frac{0.17}{\text{low}} + \frac{0.33}{\text{moderate}} + \frac{0.30}{\text{high}} + \frac{0.18}{\text{very high}}$$

$$C1 = \frac{0.02}{1} + \frac{0.17}{2} + \frac{0.33}{3} + \frac{0.30}{4} + \frac{0.18}{5}$$

The membership function is also expressed as (0.02, 0.17, 0.33, 0.30, 0.18). Similarly, the membership functions of the other sixteen PRFs and the five

Table 7.4 Membership functions of all PRFs in relation to risk severity (Chan, 2011)

PRF	W	MF of Level 3	MF of Level 2
RF 16	0.16	(0.01,0.21,0.38,0.32,0.08)	(0.03,0.16,0.34,0.35,0.12)
RF 4	0.16	(0.05,0.17,0.40,0.29,0.09)	
RF 3	0.18	(0.01,0.12,0.28,0.38,0.21)	
RF 20	0.17	(0.03,0.15,0.33,0.39,0.10)	
RF 18	0.16	(0.03,0.16,0.35,0.29,0.17)	
RF 17	0.17	(0.04,0.13,0.30,0.40,0.13)	
RF 7	0.23	(0.10,0.19,0.43,0.20,0.08)	(0.04,0.17,0.36,0.30,0.13)
RF 1	0.26	(0.02,0.17,0.33,0.30,0.18)	
RF 2	0.25	(0.02,0.18,0.38,0.33,0.09)	
RF 9	0.26	(0.04,0.13,0.30,0.37,0.16)	
RF 32	1.00	(0.08,0.16,0.27,0.37,0.12)	(0.08,0.16,0.27,0.37,0.12)
RF 5	0.34	(0.03,0.10,0.37,0.30,0.20)	(0.02,0.12,0.35,0.35,0.16)
RF 29	0.33	(0.02,0.14,0.30,0.39,0.15)	
RF 6	0.33	(0.02,0.12,0.37,0.37,0.12)	
RF 26	0.36	(0.07,0.09,0.26,0.25,0.33)	(0.06,0.13,0.32,0.31,0.18)
RF 21	0.32	(0.03,0.18,0.33,0.36,0.10)	
RF 22	0.32	(0.07,0.11,0.37,0.32,0.13)	

Notes: PRF = principal risk factor, W = weighting, MF = membership function.

Table 7.5 Membership functions of all PRFs in relation to risk likelihood (Chan, 2011)

PRF	W	MF of Level 3	MF of Level 2
RF 16	0.16	(0.01,0.13,0.27,0.34,0.14,0.09,0.02)	(0.04,0.11,0.22,0.32,
RF 4	0.16	(0.03,0.11,0.21,0.33,0.14,0.12,0.06)	0.16,0.11,0.04)
RF 3	0.18	(0.09,0.16,0.19,0.30,0.15,0.07,0.04)	
RF 20	0.17	(0.04,0.08,0.19,0.33,0.17,0.12,0.07)	
RF 18	0.16	(0.04,0.12,0.24,0.33,0.17,0.12,0.07)	
RF 17	0.17	(0.02,0.07,0.20,0.29,0.19,0.19,0.04)	
RF 7	0.23	(0.07,0.14,0.12,0.20,0.22,0.17,0.08)	(0.07,0.14,0.19,0.26,
RF 1	0.26	(0.07,0.14,0.12,0.20,0.22,0.17,0.08)	0.20,0.10,0.04)
RF 2	0.25	(0.03,0.14,0.16,0.36,0.23,0.04,0.04)	
RF 9	0.26	(0.09,0.10,0.29,0.28,0.13,0.09,0.02)	
RF 32	1.00	(0.03,0.11,0.26,0.24,0.24,0.10,0.02)	(0.03,0.11,0.26,0.24, 0.24,0.10,0.02)
RF 5	0.34	(0.03,0.11,0.26,0.24,0.24,0.10,0.02)	(0.03,0.14,0.14,0.28,
RF 29	0.33	(0.03,0.17,0.12,0.34,0.23,0.07,0.04)	0.22,0.13,0.06)
RF 6	0.33	(0.02,0.16,0.18,0.25,0.21,0.11,0.07)	
RF 26	0.36	(0.09,0.20,0.22,0.28,0.10,0.04,0.07)	(0.05,0.13,0.21,0.33,
RF 21	0.32	(0.02,0.11,0.18,0.20,0.19,0.13,0.07)	0.15,0.07,0.06)
RF 22	0.32	(0.03,0.07,0.23,0.41,0.19,0.04,0.03)	

Notes: PRF = principal risk factor, W = weighting, MF = membership function.

PRGs for both severity and likelihood are indicated in Table 7.4 and Table 7.5 respectively.

Development of a fuzzy synthetic evaluation model for TCC and GMP projects in Hong Kong

After generating the weightings for the seventeen PRFs and five PRGs, together with the fuzzy membership functions for each PRF, a total of four models were used to generate the results of the evaluation (Lo, 1999).

$$\text{Model 1: } M\ (\wedge,\ \vee)\text{: } bj = \overset{m}{\underset{i=1}{V}}(wi \wedge rij) \qquad \forall bj \in B$$

$$\text{Model 2: } M\ (\bullet,\vee)\text{: } bj = \overset{m}{\underset{i=1}{V}}(wi \times rij) \qquad \forall bj \in B$$

Both Model 1 and Model 2 are suitable for single-item problems, as only the major criteria are taken into account; other minor criteria are ignored (Lo, 1999). Due to the involvement of multi-criteria in the calculation of the Overall Risk Index, each PRF will have an influence on the overall risk level. Hence, both Models 1 and 2 are considered inappropriate for this study.

Model 3: $M\ (\bullet, \oplus): bj = \min(1, \sum_{i=1}^{m} wi \times rij)$ $\forall bj \in B$

Model 4: $M\ (\wedge, +): bj = \sum_{i=1}^{m} (wi\wedge rij)$ $\forall bj \in B$

The symbol $\oplus$ in Model 3 represents the summation of the product of the weighting and membership functions. Model 3 is appropriate when many criteria are taken into consideration and the difference in the weighting of each criterion is not large. Model 4 will neglect some information with smaller weightings, so it generates similar results to Models 1 and 2.

In conclusion, Model 3 is the most appropriate to determine the ORI and the respective risk indices of different PRGs for TCC and GMP projects among the four models, because the differences of the weightings for PRFs are not significant and many criteria (seventeen PRFs) are involved in the calculation of the ORI.

It is worth noting that there are three levels of membership functions. Level 3, Level 2 and Level 1 refer to each of the seventeen PRFs, each of the five PRGs and the ORI respectively.

ORI_A denotes the ORI of TCC and GMP construction projects in Hong Kong. W and R denote the weighting and membership function of each PRF (Level 2) respectively. The overall results of fuzzy synthetic evaluation are shown in Table 7.6.

After acquiring the membership function of Level 1, the ORI can be calculated using the following equation:

$$ORI_A = \sum_{k=1}^{m} (W \times R_K) \times L$$

where:

ORI_A is the Overall Risk Index;
W is the weighting of each PRF;
R is the degree of membership function of each PRF;
L is the linguistic variable, where 1 = very low, 2 = low, 3 = moderate, 4 = high and 5 = very high (for severity); and 1 = very very low, 2 = low, 3 = low, 4 = moderate, 5 = high, 6 = very high and 7 = very very high (for likelihood):

ORI of TCC and GMP construction projects in Hong Kong

= $(0.04 \times 1 + 0.15 \times 2 + 0.34 \times 3 + 0.33 \times 4 + 0.14 \times 5) \times (0.05 \times 1 + 0.13 \times 2 + 0.20 \times 3 + 0.30 \times 4 + 0.18 \times 5 + 0.11 \times 6 + 0.05 \times 7)$

= 3.38×4.02

= 13.59

Table 7.6 Results of fuzzy synthetic evaluation for all PRGs (Chan, 2011)

	Principal risk group	W	MF for Level 2	MF for Level 1
Risk severity (from Level 2 to Level 1)	Pre-contract risks	0.35	(0.03,0.16,0.34,0.35,0.12)	**(0.04,0.15,**
	Post-contract risks	0.23	(0.04,0.17,0.36,0.30,0.13)	**0.34,0.33,**
	Lack of experience in TCC and GMP procurement process	0.06	(0.08,0.16,0.27,0.37,0.12)	**0.14)**
	Design risks	0.18	(0.02,0.12,0.35,0.35,0.16)	
	Economic and financial risks	0.18	(0.06,0.13,0.32,0.31,0.18)	
Risk likelihood (from Level 2 to Level 1)	Pre-contract risks	0.35	(0.04,0.11,0.22,0.31,0.16, 0.11,0.04)	**(0.05,0.13,** **0.20,0.30,**
	Post-contract risks	0.23	(0.07,0.14,0.19,0.26,0.20, 0.10,0.04)	**0.18,** **0.11,0.05)**
	Lack of experience in TCC and GMP procurement process	0.06	(0.03,0.11,0.26,0.24,0.24, 0.10,0.02)	
	Design risks	0.18	(0.03,0.14,0.14,0.28,0.22, 0.13,0.06)	
	Economic and financial risks	0.18	(0.05,0.13,0.21,0.33,0.15, 0.07,0.06)	

Notes: MF = membership function; W = weighting in Table 7.3; MF of Level 1 = sum-product of weighting and MF of Level 2.

The findings established by the fuzzy synthetic evaluation showed that the ORI of TCC and GMP projects was 13.59, regarded as higher than 'moderate' as it was greater than the median value of 12 (severity of 3 multiplied by likelihood of 4). In addition, the risk index of a particular PRG could be measured by the same method in order to conduct an in-depth analysis. Table 7.7 shows the aggregate results.

As shown in Table 7.7, 'design risks' were considered the most significant risk group. The second was 'pre-contract risks', followed by 'economic and financial risks'. 'Lack of experience in TCC and GMP procurement process' ranked fourth, and 'post-contract risks' ranked last. Table 7.7 shows that design risks (shown in bold) may be a major hurdle in achieving project

Table 7.7 Risk indices of principal risk groups (Chan, 2011)

Principal risk group	Severity	Likelihood	Risk index	Rank
1. Pre-contract risks	3.37	3.94	13.28	2
2. Post-contract risks	3.31	3.84	12.71	5
3. Lack of experience in TCC and GMP procurement process	3.29	3.93	12.93	4
4. Design risks	**3.51**	**4.15**	**14.57**	**1**
5. Economic and financial risks	3.42	3.85	13.17	3
Overall risk level	**3.38**	**4.02**	**13.59**	

success under TCC and GMP methods in Hong Kong. For instance, variations can be an important risk for TCC and GMP projects. According to Tang and Lam (2003), changes in scope of work under the TCC and GMP approaches may lead to disputes (Tang and Lam, 2003). Construction projects are vulnerable to the external environment, such as changes in economic climate and market demand, leading to changes in the scope of design and work. As any unexpected changes in scope of work would probably result in a large number of TCC and GMP variations (Fan and Greenwood, 2004), they would require more time and incur considerable cost implications for the projects.

In addition, it is difficult to identify the extent of design development. Inappropriate management of these issues might cause intractable disputes and thus damage the established partnering relationship and mutual trust within the project team (Sadler, 2004). These findings could help both employer organisations and main contractors to better comprehend how the key risk factors are identified, analysed, calculated, evaluated and reduced for future TCC and GMP construction projects. Hong Kong is a good example, as it exhibits similar features (e.g. predominantly high-rise complex construction) to other metropolitan cities around the globe.

Potential applications of the FRAM

At the pre-contract stage, the employer/developer may compute the assessment of the seventeen PRFs into the model to develop a risk index for using TCC and GMP schemes individually, then to decide whether to adopt TCC or GMP or neither of them for a prospective construction project.

At tender stage, the contractor could utilise the FRAM to evaluate the risk level of the project procured by TCC and GMP approaches during the peer review stage in bidding and determine whether to bid or not. By considering the Overall Risk Index, the builder could quantify the risk exposure of the projects and enhance the pricing process during preparation of the tender.

At the post-contract stage, the project stakeholders could compute the seventeen PRFs to establish an Overall Risk Index in order to evaluate the risk level periodically, such as at every monthly meeting for monitoring the risks. The most critical PRG could be identified by the ORI, so that stakeholders could focus on the constituent component risks of the PRG. Hence, the overall risk level could be reduced.

In addition, the priority of the risk groups could be arranged to manage the risks based on their corresponding risk indices. For instance, if the risk index of design risks is the highest among five PRGs, the project team could manage the risks. In this way, the model can show the priorities of risk mitigation/management.

Generally, the model has developed a norm for TCC and GMP projects in Hong Kong. If there are sufficient TCC and GMP samples/projects, a benchmarking model can be established to help project stakeholders to benchmark the risk level: say, the Overall Risk Index is 80% of the norm (i.e. lower than the norm) or 120% of the norm (i.e. higher than the norm).

As the methodology of this model building is suitable for every geographical region, it could be adopted in other regions employing TCC and GMP contracts to enable international comparisons of risk levels of TCC and GMP projects among countries/regions. Similarly, this methodology can be applied to evaluate and compare the risk level of construction projects under traditional contracts with the risk level of TCC and GMP projects.

For validation of the risk assessment model, refer to Chan (2011).

Chapter summary

This chapter has demonstrated the development of an objective and comprehensive risk assessment model for TCC and GMP projects, together with the concept of fuzzy synthetic evaluation model. The introduction of this model can improve the perceptions of project team stakeholders regarding achieving successful TCC and GMP projects. It can also provide a platform to evaluate the risks of the projects by considering objective evidence instead of intuitive judgements. The study found that design risk is the most critical risk group encountered in TCC and GMP construction projects. This might be attributable to the grey areas when considering whether a variation should be regarded as a design development variation or a contract variation, involving cost implications for the projects.

Subsequent to the establishment of the model, a validation exercise was conducted via face-to-face interviews in 2010 with seven experts who had direct hands-on experience in TCC and GMP projects in Hong Kong. They rated the comprehensiveness of the risks identified in the model, along with its clarity, objectivity, practicality and overall reliability. The ratings were found to be satisfactory in all aspects in this validation exercise. Therefore, the model was perceived as comprehensive, clear, objective, practical and reliable by the participants in the validation process.

8 Risk allocation in target cost contracts

Introduction

Not all TCC and GMP projects are equally successful in terms of time performance, cost performance and quality performance, as clients conventionally use exculpatory contractual provisions to reduce or avoid their own responsibilities in the contracts. In the long run, this onerous allocation of risks may not be in the interest of the construction industry. Some employers share the risk arbitrarily (Mosey, 2009). In fact, both the employer and contractor are betting against the accuracy of the risk priced. Transferring risks as far as possible may bring short-term benefits, but it may lead to an adversarial atmosphere, reluctance to tender for future works, and intractable disputes (Zaghloul and Hartman, 2003).

This research aims to help industrial practitioners and researchers explore suitable risk allocation for TCC and GMP arrangements, and provide some fundamentals for future research (e.g. international comparison of various risk allocation choices in TCC and GMP contracts).

Li et al. (2005) define risk allocation as primary measures of risk assignment among project stakeholders. When risk outcomes are allocated to contractual parties, there will be a risk allocation mechanism.

The risk allocation may be stipulated in the contractual language and exculpatory clauses (e.g. 'no damage for delay' clauses and indemnification clauses) which are usually utilised to transfer one's risk in common law to another stakeholder. Nevertheless, Ashley et al. (1989) were sceptical of their enforceability, as even if the exculpatory clauses are enforceable, contractors tend to inflate their tender prices in order to transfer the cost to the employer. Ahmed et al. (1998) echo that misallocation of risk may lead to higher costs borne by employers because of bid contingency. Therefore, it is important to allocate the risk impartially and evenly in order to achieve project success.

Citing Ward et al. (1991), Edwards (1995) and Abednego and Ogunlana (2006), Flanagan and Norman (1993) found that the following five conditions should be fulfilled to determine whether the risks are shared properly:

1 Risks should be transferred to the party which will be capable of handling them.
2 Risk should be clearly identified, comprehended and evaluated by stakeholders.
3 A party must be able to handle the risks in terms of management and technology.
4 A party must have sufficient capital to bear the consequence of the risk and/or mitigate the risks.
5 A party must be willing to accept the risks.

Risk allocation

J.H.L. Chan et al. (2011a) have documented the research findings reported in this section. These findings were modified based on El-Sayegh (2008), who investigated risk assessment and risk allocation in the United Arab Emirates. According to the study of El-Sayegh (2008), the survey mainly focused on the suitable allocation of forty-two identified risks. There was a similar study which invited industrial practitioners to determine the best capable party (employer and/or contractor) to handle the risks in TCC and GMP construction projects based on their past experience under TCC and GMP arrangements. According to Cooper et al. (2005), a general principle is to allocate each risk to the party which can best manage the risk with the minimum cost. Similarly, according to Ke et al. (2010), a desirable risk allocation should not be transfer all risks to one stakeholder, but should find a suitable solution to minimise the overall management costs among the employer and contractors.

Table 8.1 shows the meaning of each choice relating to the fraction of risk allocation between employer and contractor. J.H.L. Chan et al. (2011b) conducted a questionnaire survey to identify the appropriate risk allocation in TCC. The participants in the survey were required to select the 'party best capable to manage the risk' corresponding to each of the thirty-four risk factors in the scale in Table 8.3.

Table 8.1 Meanings of choices in the survey (J.H.L. Chan et al., 2011b) (with permission from Emerald Group Publishing Ltd)

1	Client (100%)	Client is best capable to manage the risk
2	Client > contractor	Client is more capable than contractor to manage the risk
3	Client = contractor	Both client and contractor are equally capable to manage the risk
4	Contractor > client	Contractor is more capable than client to manage the risk
5	Contractor (100%)	Contractor is best capable to manage the risk

Table 8.2 Interpretation of survey findings (J.H.L. Chan et al., 2011b) (with permission from Emerald Group Publishing Ltd)

Case	Result	Party perceived as best capable to manage the risk
1	Total percentage of Choice 1 plus Choice 2 > 50%	Client
2	Total percentage of Choice 4 plus Choice 5 > 50%	Contractor
3	Percentage of Choice 3 > 50%	Shared
4	None of Cases 1–3	Negotiated

The 'perceived party best capable to manage the risk' will be established if the vote for such a risk is more than 50%, as in the studies of El-Sayegh (2008) and Li et al. (2005). As the party best suited to managing the risk is better able to bear it, the interpretation of findings is shown in Table 8.2.

Risks to be allocated to client

Table 8.3 depicts eight risks best allocated to the employer:

- change in scope of work;
- errors and omissions in tender document;
- inaccurate topographical data at tender stage;
- insufficient design completion during tender invitation;
- poor buildability/constructability of project design;
- lack of involvement of main contractor in design development process;
- unforeseeable design development risks at tender stage;
- consequence of delayed payment to contractor.

The above risks can be divided into three categories: contractual risks, design risks, and economic and financial risks.

'Change in scope of work', 'errors and omissions in tender document' and 'inaccurate topographical data at tender stage' could be categorised as contractual risks. 'Change in scope of work' was perceived as one of the important risks under TCC and GMP procurements. Research in the United Kingdom by Cox et al. (1999) found that alteration of the client's requirements was one of major causes for design changes. This risk is thus better transferred to the employer. The result is in line with the finding of Ojo and Ogunsemi (2009) that the risk of 'change in work' is regarded as being borne by employer. Another two risks relate to tender preparation. The majority of respondents felt that these risks have to be transferred to employer. A possible reason may be that the above-mentioned three risks are controlled by the employer. Laryea (2011) conducted case studies in

Table 8.3 Preferred allocation of risk factors in TCC and GMP construction projects in Hong Kong (J.H.L. Chan et al., 2011b) (with permission from Emerald Group Publishing Ltd)

No.	Risk factor	Risk allocation			
		Client	Shared	Contractor	Allocated to
5	Change in scope of work	80.9%	13.8%	5.3%	Client
6	Errors and omissions in tender document	64.5%	15.1%	20.4%	Client
8	Inaccurate topographical data at tender stage	61.3%	22.6%	16.1%	Client
17	Insufficient design completion during tender invitation	79.6%	16.1%	4.3%	Client
18	Poor buildability/constructability of project design	50.5%	22.6%	26.9%	Client
19	Lack of involvement of main contractor in design development process	68.8%	12.9%	18.3%	Client
20	Unforeseeable design development risks at tender stage	65.6%	24.7%	9.7%	Client
25	Consequence of delayed payment to contractor	73.4%	18.1%	8.5%	Client
7	Difficult for main contractor to have back-to-back TCC and GMP contract terms with nominated or domestic subcontractors	8.7%	13.0%	78.3%	Contractor
12	Poor quality of work	6.5%	17.2%	76.3%	Contractor
13	Delay in availability of labour, materials and equipment	2.1%	17.0%	80.9%	Contractor
14	Low productivity of labour and equipment	1.1%	10.6%	88.3%	Contractor
15	Selection of subcontractors with unsatisfactory performance	4.3%	23.4%	72.3%	Contractor
24	Change in interest rate on main contractor's working capital	5.4%	24.7%	69.9%	Contractor
2	Delay in resolving contractual disputes	25.8%	64.5%	9.7%	Shared
4	Disagreement over evaluating the revised contract price after submitting an alternative design by main contractor	26.9%	57.0%	16.1%	Shared
10	Difficult to agree on a sharing fraction of saving/overrun of budget at pre-contract award stage	16.1%	77.4%	6.5%	Shared
22	Inflation beyond expectation	19.1%	51.1%	30.9%	Shared

(continued)

Table 8.3 (continued)

No.	Risk factor	Risk allocation			
		Client	Shared	Contractor	Allocated to
27	*Force majeure* (Acts of God)	10.8%	78.5%	10.7%	Shared
28	Inclement weather	7.5%	57.0%	35.5%	Shared
30	Change in relevant government regulations	35.5%	60.2%	4.3%	Shared
32	Lack of experience of contracting parties throughout TCC and GMP process	20.4%	59.1%	20.5%	Shared
1	Actual quantities of work required far exceeding estimate	41.3%	32.6%	26.1%	Negotiated
3	Unrealistic maximum price or target cost agreed in the contract	38.3%	41.5%	20.2%	Negotiated
9	Loss incurred by main contractor due to unclear scope of work	45.2%	29.0%	25.8%	Negotiated
11	Technical complexity and design innovations requiring new construction methods and materials from main contractor	12.8%	41.5%	45.7%	Negotiated
16	Delay in work due to third party	23.4%	44.7%	31.9%	Negotiated
21	Exchange rate variations	18.1%	42.6%	39.3%	Negotiated
23	Market risk due to the mismatch of prevailing demand of real estate	45.7%	41.5%	12.8%	Negotiated
29	Unforeseeable ground conditions	32.6%	42.4%	25.0%	Negotiated
31	Difficult to obtain statutory approval for alternative cost-saving designs	29.8%	37.2%	33.0%	Negotiated
33	Impact of construction project on surrounding environment	17.2%	44.1%	38.7%	Negotiated
34	Environmental hazards of constructed facilities towards the community	24.5%	44.7%	30.8%	Negotiated

the United Kingdom and found that the four major reasons for deteriorated quality of tender documents were employer impatience, reluctance to invest more in improving the quality of tender documents, negligence and incompetence. Moreover, the employer is likely to provide inaccurate topographical information to the contractor at tender stage. Although in most cases employers do not guarantee the accuracy of topographical data, they have full control over this risk.

It is recommended that design risks, including 'insufficient design completion during tender invitation', 'poor buildability/constructability of project design', 'little involvement of main contractor in design development process' and 'unforeseeable design development risks at tender stage', be borne by the employer. The result is comprehensible and echoes findings from the previous literature (Kartam and Kartam, 2001; Andi, 2006). The design consultants, such as architects, structural engineers and building services engineers, usually undertake the entire design work because of their professional training and inherent expertise, representing the employer's intent and interests. However, under traditional procurement practices in the Hong Kong construction industry, the contractor is passive in terms of changes of design, so the employer will be better able to handle these design risks.

'Consequence of delayed payment to contractor' was the last risk which had to be transferred to the employer. This is in line with the study by Andi (2006). The standard form of contract (e.g. NEC3 Option C) stipulates that interest will be paid on late payment if a certified payment is late due to delays in issuing the certificate by the project manager.

It is not complicated to observe that the risks which are better transferred to the employer are all controlled by the employer, such as change in scope of works, late payment, as well as errors and omissions in tender document. Grove (2000) concurred with the results regarding inappropriate standards and management standards. Based on the fault standard, time and cost effects due to risks induced by faults of a particular party have to be allocated to that party. It is apparent that faults of the employer lead to such risks, hence this survey finding matched the fault standard. However, as mentioned by Grove (2000), the rationale of the management standard is to allocate risks to the party which can manage, estimate and handle them. The employer may have control over all risks stated in this section, such as insufficient design completion at tender invitation stage, and change in scope of work. Therefore, the findings were considered sensible and representative of real-life circumstances.

Risks to be allocated to contractor

As shown in Table 8.3, six risks are better allocated to and managed by contractors:

- difficult for main contractor to have back-to-back TCC and GMP contract terms with nominated or domestic subcontractors;
- poor quality of work;

- delay in availability of labour, materials and equipment;
- low productivity of labour and equipment;
- selection of subcontractors with unsatisfactory performance;
- change in interest rate on main contractor's working capital.

In respect of site operation, 'poor quality of work', 'delay in availability of labour, materials and equipment', 'low productivity of labour and equipment', and 'selection of subcontractors with unsatisfactory performance' are risks that are better transferred to the builder. Around 70% of the survey participants opined that these risks are better handled by contractor under TCC and GMP arrangements, as shown in Table 8.2.

The result is reasonable, as the contractor is the actual constructor. Contractors are better able to control the risks in site operation. According to the findings from the fuzzy risk allocation model in the study of Lam et al. (2007), the contractor is in a better position to manage the risk of unsatisfactory quality control of work by subcontractors. Andi (2006) also suggested that the contractor should bear the risk of late availability of labour, materials and equipment, poor quality of work, and choice of subcontractors with undesirable performance. The survey results matched the findings of Andi (2006).

About 75% of the respondents believed that 'difficult for main contractor to have back-to-back TCC and GMP contract terms with nominated or domestic subcontractors' was a risk that was more suitable to be handled by the contractor. Around 70% of the survey participants regarded that 'change in interest rate on main contractor's working capital' need to be controlled by the builder. The results were also consistent with the management standard of risk allocation of Grove (2000). The main contractor is the only stakeholder which can have full control over the management of subcontractors and contractual arrangements with them. If the interest rate on the main contractor's working capital changes, the contractor will suffer the loss if the risk does materialise. Such a risk falls within one of the principles of Abrahamson (1984). Regarding the risk of termination of contract because of the client's bankruptcy, in Hong Kong, as pointed out by Chan et al. (2010b), the employers using TCC and GMP arrangements were the related works departments of the HKSAR Government, quasi-governmental organisations, major private property developers and large-scale construction contractors. The risk of bankruptcy of such employers might not be significant. Nevertheless, the circumstances might not be the same around the globe. If the employer goes bankrupt, the contractor is required to bear the risk. Usually there are some contractual mechanisms managing this risk in Hong Kong. Non-payment by the employer will be treated as a debt in case of the client's bankruptcy.

Risks to be shared between client and contractor

There are nine risks which are perceived to be shared between the employer and contractor. By examining these nine risks closely, they can be sub-divided into two categories: (1) risks out of control of employers and contractors, and (2) risks potentially due to both parties. Risks in category 1 involve:

- 'inflation beyond expectation';
- 'global financial crisis';
- '*force majeure* (Act of God)';
- 'inclement weather';
- 'change in relevant government regulations'.

'Inflation beyond expectation', '*force majeure* (Act of God)', 'inclement weather' and 'change in relevant government regulations' are in line with the risk/obligation allocation model studied in the *No Dispute* report published in Australia (National Building and Construction Council, 1989). This indicates that the results are generally logical and sensible. Moreover, Lam et al. (2007) proposed a fuzzy risk allocation model, suggesting that inclement weather and risks of inflation should be shared between the client and the builder. In most contracts in Hong Kong, inclement weather is one of the grounds for claiming extension of time. On the other hand, contractors should assume any cost implications for this risk. Regarding the inflation risk, it is shared if there is a fluctuation clause in TCC and GMP contracts. The use of such contractual clauses may be because both parties could not manage the degree of significance and possibility of occurrence these risks. Since it will not be impartial to transfer these risks to one party, the risks are deemed to be shared between employer and builder.

There are some risks that may be contributed by both employer and contractor, namely 'delay in resolving contractual disputes', 'disagreement over evaluating the revised contract price after submitting an alternative design by main contractor' and 'lack of experience of contracting parties throughout TCC and GMP process'. For instance, 'delay in resolving contractual disputes' might be attributed to both parties when preparing and/or assessing claims. Andi (2006) also perceived this risk as a shared risk.

In respect of 'difficult to agree on a sharing fraction of saving/overrun of budget at pre-contract award stage', the proportion of risks is subject to negotiation if negotiated tendering is adopted in TCC and GMP arrangements. Chan et al. (2007b) commented that inexperienced employers and builders might damage TCC and GMP procurements. In Hong Kong, this risk may be unavoidable because there are limited projects completed under TCC and GMP procurement. The employer may decide whether to adopt

TCC and GMP procurement methods in the future project, while the builder may choose whether or not to bid for projects using TCC and GMP procurement strategies.

Chapter summary

Without adequate consideration of risk allocation, the pre-agreed objectives may not be achieved. This chapter considered the previous literature and the empirical questionnaire survey conducted by J.H.L. Chan et al. (2011b). It may be concluded that the recommendation is that risks under the client's control (e.g. risks concerning project design and tender documentation) are best assumed by the employer and construction risks are better be borne by the builder. The results from J.H.L. Chan et al. (2011b) may benefit future TCC and GMP projects as they are prevalently adopted around the globe (Walker et al., 2002; Rojas and Kell, 2008; Chan et al., 2008; Bogus et al., 2010; Chan et al., 2010d).

9 Risk mitigation in target cost contracts

Introduction

The conventional adversarial relationships between clients and builders are often attributable to their individual interests and the nature of their businesses rather than the whole project itself (Chan et al., 2012). Although there have been many projects that have adopted TCC and GMP (Trench, 1991; Walker et al., 2000), discrepancies in management systems, unfamiliarity with the procurement arrangements of major project parties and individual cultural background made TCC and GMP contracts problematic. For instance, Bogus et al. (2010) found that GMP arrangements were less likely to have cost and time growth when compared with lump-sum contracts in the United States. Nevertheless, Roja and Kell (2008) investigated GMP projects in the north-western United States and found that 75% of public school projects and 80% of non-school projects exceeded the contract GMP value. This result showed that GMP was not actually 'guaranteed'. Therefore, this chapter aims to identify methods to mitigate the risks potentially affecting overall performance of TCC and GMP projects.

Key risk mitigation measures

Chan et al. (2012) conducted an industry-wide questionnaire survey of individual construction professionals and key project stakeholders to illustrate the significance of eighteen listed risk mitigation measures. Table 9.1 shows the descending ranking, mean and standard deviation.

The most significant effective risk mitigation measure for TCC and GMP projects was 'right selection of project team'. Chan et al. (2010b) recommended that a competent project team was of paramount importance to the success of TCC projects, since inexperienced or claim-conscious builders may imperil the smooth TCC and GMP procurement process. Gander and Hemsley (1997) echoed that the selection of an experienced project team brings success of TCC and GMP construction projects, because an inexperienced team may lack clarity regarding obligations and duties. The employer needs to establish a project team which is open-minded in adapting innovative ideas. In particular, the main contractor needs to communicate proactively

Table 9.1 Results of the overall ranking of risk mitigation measures for TCC and GMP (Chan et al., 2012)

Risk mitigation measures for TCC and GMP	Frequency	Mean	Standard deviation	Rank
Right selection of project team	94	3.90	0.843	1
Mutual trust between the parties to the contract	94	3.73	1.109	2
Clearly defined scope of works in client's project brief	94	3.67	1.010	3
Early involvement of the main contractor in design development process	94	3.64	0.960	4
Proactive participation by the main contractor throughout the TCC and GMP process	94	3.61	0.895	5
Prompt valuation and agreement on any variations as they are introduced	94	3.60	0.872	6
Reasonable sharing mechanism of cost saving/overrun of budget between client and contractor	94	3.59	0.999	7
Confirming a contract GMP value or target cost after design documents are substantially completed	94	3.56	0.887	8
Sufficient time given to interested contractors to submit their bids for consideration	94	3.54	0.991	9
Tender interviews and tender briefings to ensure tenderers gain a clear understanding of scope of works involved and necessary obligations to be taken in the project	94	3.48	0.864	10
Clearly stated circumstances in which agreed GMP value or target cost can be adjusted in contracts	94	3.46	0.980	11
Establishment of adjudication committee and meetings to resolve potential disputed issues	93	3.27	0.946	12
Open-book accounting regime provided by main contractors in support of their tender pricing	93	3.24	1.136	13
Proper risk register with responsible parties assigned and agreed	94	3.23	0.977	14
Implementation of relational contracting within the project team	92	3.14	1.033	15
Development of standard contract clauses in connection with TCC and GMP schemes or methodology	94	3.04	1.004	16
Application of price fluctuation clause in the contract	94	2.90	0.928	17
Employing a third party to review the project design in compliance with prevailing building regulations and buildability at tender stage	94	2.64	0.937	18
Number (N)	**94**			

and willingly with the other major parties on the basis of pre-agreed partnering arrangements. Gander and Hemsley found that the main contractors and subcontractors procured through pre-qualification functioned well in this case study. Chan et al. (2010b) investigated the underground railway station modification works with a pre-qualification exercise which was under TCC arrangements in Hong Kong.

An appropriate project team can be selected through pre-qualification of contractors. Eriksson et al. (2009) carried out an eighteen-month longitudinal case study in Sweden investigating the methods by which construction employers could overcome barriers to the partnering approach by utilising purposeful procurement procedures. They found that the main contractors and subcontractors procured through pre-qualification functioned well in this case study.

The second most significant risk mitigation measure was 'mutual trust between the parties to the contract'. Chan et al. (2007a) showed that partnering concepts are instilled during TCC and GMP projects. Wong (2006) suggested that since TCC is usually adopted for high-risk projects, mutual trust between the client and the contractor would be needed to deal with the risks. Moreover, due to the unique methodology of the TCC approach based on jointly pre-determined and pre-agreed allocation of key risks between the employer and contractor, the employer comprehended the significance of realistic target cost estimates, which involve sufficient risk contingencies under the pain-share/gain-share methodology (Chan et al., 2010b). Therefore, mutual trust and a close working relationship are crucial in order to manage and reduce potential risks in a partnership culture.

Hartman (2000) devised a trust model enabling a detailed perception of the concepts of trust. This model identified three types of trust – competence trust, integrity trust and intuitive trust – which help contractual parties to trust each other during construction. Competence trust could be acquired by providing observable proofs, such as hands-on experience and track records of similar projects (Zaghloul and Hartman, 2003) According to Wong and Cheung (2004), integrity trust could be gained through the willingness of a contractual party to protect another party throughout construction. Wong and Cheung (2004) also found that intuitive trust was understanding which was not easily influenced by the immediate performance of the stakeholders, but could be influenced by the long-term relationships among the parties. Khalfan et al. (2007) suggested some methods to build up trust in the construction industry. They recommended that frequent and effective communications with practical actions and outcomes could enhance it. When a party consistently demonstrated reliability, trust was earned. Kadefors (2004) suggested that soft objectives emphasising relations and work procedures could be set out in construction projects with a partnering approach to mitigate the adverse effects of contract terms on the behaviours of contract stakeholders. As project partnering is common in Hong Kong TCC and GMP projects (Chan et al., 2007a), the aforementioned strategies could be applied to construction projects under TCC and GMP procurement methods.

The third most important risk mitigation measure was 'clearly defined scope of works in client's project brief'. In the survey conducted by Chan et al. (2010d), 'change in scope of works' was the most obvious, it can thus be deduced that respondents realized the importance of well-defined scope of works, which can minimize the risks associated with TCC/GMP projects in the construction stage.

Table 9.1 shows that if the scope of works can be clearly defined when the project starts, the risks induced by TCC and GMP measures during construction can be mitigated. This outcome echoes the study in the United Kingdom by Olawale and Sun (2010), who found that clear differentiation between a design change and a design development at the beginning of the project could reduce the risks arising from future design changes. According to Gander and Hemsley (1997), since design development is constantly evolving in TCC and GMP projects, different interpretations of whether variations should be considered as TCC and GMP variations or design development can result in disputes (Gander and Hemsley, 1997). Therefore, defining the scope of work accurately and clearly is crucial at the commencement of the project in order to minimise scope changes or other variations.

Holding a briefing session at the initial stage of project development may help to define a more detailed and accurate scope of works. Briefing in the early design stages is important for projects to achieve success, as the employer's intensions and requirements can be identified and articulated (Yu et al., 2007). The lack of a comprehensive framework for listing the needs of clients may be the result of inadequate briefing. A comprehensive investigation by Shen and Chung (2006) to explore the use of briefing in the local construction industry recommended that information technology and value management can help to identify the employer's requirements in the briefing.

The fourth most efficient way to reduce risks under TCC and GMP contracts is 'early involvement of the main contractor in design development process'. Song et al. (2009) expressed that this measure aims to engage contractors in the design process in order to enlist their construction competence in the design process. Mosey (2009) expressed that it is common that the design is not formed by design consultants only, but also includes the main contractors and specialist contractors in order to achieve a thorough and practical design. A case study of early contractor participation conducted by Song et al. (2009) showed that the benefits brought by early builder involvement involve better quality of drawings, prompt materials supply and quick information flow. The case study also concluded that if contractors are involved earlier, the project duration will be reduced due to the better design and utilisation of their experience and techniques. This result is in line with risk reduction in TCC and GMP contracts, because builders' techniques and experience in both construction methods and design can be capitalised to improve the constructability of the intended design in TCC and GMP contracts (Chan et al., 2010b). Consequently, early participation of contractors in the initial stages of

TCC and GMP contracts can reduce the potential for design changes after awarding the contracts.

The fifth most obvious risk mitigation measure is 'proactive participation of the main contractor throughout the TCC and GMP process'. There is an early warning clause in NEC3 Option C (and D) which stipulates proactive involvement of the project manager and builder to provide early warning to the contracting parties about any matters which might lead to an increase in the overall cost, late completion, and late fulfilment of major milestone dates and/or reduce the quality of performance. A risk mitigation meeting provides an opportunity for project members to develop plausible solutions in order to reduce the adverse effects of potential risks. This mechanism has been considered an efficient way to mitigate risk in NEC3 by developing a suitable risk register and involving responsible project members to assign and agree it. Furthermore, Bayliss et al. (2004) showed that team building activities, such as mutually agreed project goals and a shared site office, can help to create shared values among project stakeholders. Similarly, Eriksson et al. (2009) found that the involvement and commitment of all major contract stakeholders could increase value creation, bringing benefits to the project performance of a complex construction project of a manufacturing plant for Swedish pharmaceutical products.

Factor analysis of risk mitigation measures

The methodology of factor analysis can be referred to Chapter 7. Table 9.2, based on Chan et al. (2012), shows the grouped risk mitigation measures investigated in descending order of importance, to identify their underlying characteristics.

Interpretation of the underlying grouped risk mitigation measures

Factor 1: relational contracting and mutual trust

Factor 1 is comprised of four components primarily emphasising relational contracting and mutual trust among contracting stakeholders. As illustrated in Table 9.2, the loadings on Factor 1 were relatively high compared to the other factors. Factor 1 includes 'implementation of relational contracting within the project team', 'open-book accounting regime provided by main contractors in support of their tender pricing', 'sufficient time given to interested contractors to submit their bids for consideration' and 'mutual trust between the parties to the contract'. The common features of these items are related to the underlying relationship among project stakeholders. Mutual trust and contracting strategy are inextricably linked, and they are crucial to achieve effective contract administration and project management (Zaghloul and Hartman, 2003). Tay et al. (2000) suggested that a close relationship among all the project stakeholders is significant in the successful implementation of TCC. As summarised by Chan et al. (2007b), partnering

Table 9.2 Results of factor analysis of the eighteen risk mitigation measures for TCC and GMP schemes (Chan et al., 2012)

No.	Item	Factor loading	Eigenvalue	% of variance explained	Cumulative % of variance explained
Factor 1 – Relational contracting and mutual trust					
10	Implementation of relational contracting within the project team	0.828	4.661	25.893	25.893
13	Open-book accounting regime provided by main contractors in support of their tender pricing	0.725			
11	Sufficient time given to interested contractors to submit their bids for consideration	0.662			
12	Mutual trust between the parties to the contract	0.591			
Factor 2 – Clear contract provisions and scope of works					
2	Clearly stated circumstances in which agreed GMP value or target cost can be adjusted in contracts	0.771	2.003	11.127	37.020
1	Application of price fluctuation clause in the contract	0.671			
3	Clearly defined scope of works in client's project brief	0.662			
6	Confirming a contract GMP value or target cost after design documents are substantially completed	0.661			
Factor 3 – Involvement of contractor in decision making process					
18	Establishment of adjudication committee and meetings to resolve potential disputed issues	0.754	1.449	8.047	45.067
15	Reasonable sharing mechanism of cost saving/overrun of budget between client and contractor	0.730			
8	Early involvement of the main contractor in design development process	0.709			

Factor 4 – Right selection of project team

16	Right selection of project team	0.853	1.337	52.497
14	Proactive participation by the main contractor throughout the TCC and GMP process	0.808	7.430	
5	Proper risk register with responsible parties assigned and agreed	0.556		

Factor 5 – Third-party review of project design at tender stage

9	Employing a third party to review the project design in compliance with prevailing building regulations and buildability at tender stage	0.801	1.132	58.786
			6.290	

Factor 6 – Standard contract clauses for TCC and GMP schemes

7	Development of standard contract clauses in connection with TCC and GMP schemes or methodology	0.701	1.054	64.639
			5.853	

Factor 7 – Fair treatment of contractor

4	Prompt valuation and agreement on any variations as they are introduced	0.833	1.002	70.208
17	Tender interviews and tender briefings to ensure tenderers gain a clear understanding of scope of works involved and necessary obligations to be taken in the project	0.653	5.569	

can be applied with TCC and GMP procurement methods to achieve project success. Partnering, which is a type of relational contracting, can promote mutual trust, enhance communication flow, improve working relationships and expedite the resolution of disputes among major stakeholders (Chan et al., 2004a). Thus, the implementation of relational contracting together with mutual trust among contracting parties can mitigate the risks associated with TCC and GMP projects, which usually pertain to design changes and scope of works, resulting in better information flows and working relationships among project stakeholders.

Factor 2: clear contract provisions and scope of works

Factor 2 involves four items regarding tender and contract documents. Chan et al. (2010d) found that the significant risk factors incurred by TCC and GMP contracts are changing scope of works, nature of variations and clarifying tender documents in the local construction industry. In response to such risks, including detailed and accurate contract clauses and scope of works in the employer's project brief is likely to decrease the risk of disputes concerning the nature of changes and scope of works. It is recommended to identify situations which could modify the pre-agreed target cost or GMP value in contracts in order to reduce the potential for claims and disputes after awarding the contract (Fan and Greenwood, 2004).

Factor 3: involvement of contractor in decision making process

Factor 3 consists of three items concerning the participation of the contractor at the decision making stage: 'establishment of adjudication committee and meetings to resolve potential disputed issues', 'reasonable sharing mechanism of cost saving/overrun of budget between client and contractor' and 'early involvement of the main contractor in design development process'. Chan et al. (2010b) observed that TCC and GMP schemes together with a partnering spirit encourage deeper collaboration between the client and the main contractor. A platform for overcoming difficulties and resolving disputes can be established through regular partnering review meetings and the adjudication committee. This result echoes the investigation by Rose and Manley (2010) concerning financial incentive mechanisms in Australia. They also found that design and construction could be improved by capitalising the builder's skills to increase buildability. Sidwell and Kennedy (2004) also suggested that participation of the contractor at an early stage can reduce uncertainty during construction.

Factor 4: right selection of project team

Factor 4 covers three items: 'right selection of project team', 'proactive participation by the main contractor throughout the TCC and GMP process' and 'proper risk register with responsible parties assigned and agreed'. A case study of an underground railway station modification project reported by Avery (2006) indicated that having a suitable project team could enhance mutual trust and effective communications among members. To deal with unforeseen matters and possible disputes, proactive builders and strong leadership are essential. The decisions of any member could make or mar the project strategy. Therefore, Factor 4 is crucial to decrease the risks brought by inexperienced project members.

Factor 5: third-party review of project design at tender stage

Factor 5 involves only one item: 'employing a third party to review the project design in compliance with prevailing building regulations and buildability at tender stage'. The client can review the design prior to tender documentation, and thus minimise the amount of errors and omissions in both the tender and contract documents. One of the risk mitigation measures for design changes in construction projects is to appoint a design manager to oversee design changes and review related information. Chan et al. (2010d) carried out seven in-depth interviews with construction professionals with abundant hands-on experience in TCC and GMP construction projects, demonstrating that one of the risk mitigation measures for TCC and GMP is to conduct a third-party review of the project design to comply with current building regulations and buildability at the tender stage.

Factor 6: standard contract clauses for TCC and GMP schemes

Factor 6 covers only one item. As stated in Chan et al. (2007a), standard contracts for TCC and GMP procurements are necessary to ensure the success of TCC and GMP projects. Although NEC3 engineering and construction contracts have been implemented in recent decades (including Option C – Target Cost with Activity Schedule, and Option D – Target Cost with Bills of Quantities), they are still limited in Hong Kong. Before 2012, only one case under NEC3 Option C could be explored (Cheung, 2008). Regarding GMP contracts, Chen et al. (2007a) stated that developers are prone to use their in-house standard contracts and modify them to adapt to the GMP strategy. A standard form of GMP contract in Hong Kong is essential to improve the receptivity of this procurement method (Ting, 2006).

Factor 7: fair treatment of contractor

Regarding Factor 7, there are two items concerning fair treatment of builders: 'prompt valuation and agreement on any variations as they are introduced' and 'tender interviews and tender briefings to ensure tenderers gain a clear understanding of scope of works involved and necessary obligations to be taken in the project'. Establishing a clear perception of objectives or achievements at the beginning of the project is crucial (Bower et al., 2002). Therefore, it is recommended to hold a tender interview and briefing session to ensure that potential tenderers clearly understand the scope of works and operational mechanism under TCC and GMP. During tender interviews, tenderers can establish the possible risks before awarding the contract. Tender briefings can be transparent, comprehensive and impartial to all interested parties. Timely valuation of variations can reduce the possibility of disputes and intractable claims regarding the quantity and nature of variations. If it is impossible to reach agreement on valuations, the project members can use the dispute resolution mechanism stated in the contract to reduce the adverse effects of other construction activities in the construction stage. The two items mentioned are impartial towards both employer and contractor in minimising potential disputes or claims.

Chapter summary

The five most efficient risk mitigation measures suggested by industrial practitioners involve: (1) 'right selection of project team', (2) 'mutual trust between the parties to the contract', (3) 'clearly defined scope of works in client's project brief', (4) 'early involvement of the main contractor in design development process' and (5) 'proactive participation by the main contractor in the delivery of TCC and GMP projects'. Moreover, Chan et al. (2012) found that the risk mitigation measures focused on two areas:

1   relationship management, such as 'relational contracting and mutual trust' and 'involvement of contractor in decision making process';
2   the tendering process, such as 'clear contract provisions and scope of works', 'third-party review of project design at tender stage' and 'standard contract clauses for GMP/TCC schemes'.

The result was reasonable, as the success of TCC and GMP projects relies heavily on a partnering spirit and the definition of a clear scope of works at the beginning of the project (Chan et al., 2010b).

With the above risk mitigation measures, decision makers can determine whether to use TCC and GMP procurement methods in their own projects. In addition, industrial practitioners have generated many useful and practical strategies for reducing the potential risks. It is expected that there will be more TCC and GMP applications in the entire construction industry.

This chapter has covered some fundamental risk mitigation solutions which are considered to be appropriate for high-risk projects (Wong, 2006).

Part IIB

Performance measurement in target cost contracts

10 A conceptual framework for identifying key performance indicators for target cost contracts

Introduction

There are diverse opinions on the benefits brought about by TCC and GMP procurement forms. Hughes et al (2011) opined that TCC might not motivate the contractor to minimise the cost. On the other hand, Chan et al (2010a) conducted eight face-to-face interviews and found that TCC and GMP procurement forms could provide financial motivation for the contractor to save cost and innovate. The performance outcomes of TCC and GMP construction projects would thus be an interesting topic of research to provide an objective tool for measuring the overall success of those projects. Based on a research study by Chan and Chan (2012), a conceptual framework for identifying the Key Performance Indicators (KPIs) for TCC is introduced in this chapter.

Performance of TCC and GMP projects

United Kingdom

The New Wembley Stadium in London, under GMP procurement, was opened in March 2007 (Mylius, 2007). The final cost of this project was more than £757 million, which significantly exceeds the original budget (£200 million) in 1996. The project was delayed by almost two years. Meng and Gallagher (2012) examined sixty completed projects in relation to cost, time and quality performance in the United Kingdom and the Republic of Ireland. In respect of cost certainty, fixed-price contracts were more satisfactory than TCC. They found that 70% of projects procured by fixed-price contracts in the study were completed on budget or achieved cost savings, while only about half of projects under the TCC arrangement were accomplished within or under budget.

Australia

Hauck et al. (2004) studied a project of the National Museum of Australia which was procured with the TCC strategy, and found that it achieved outstanding project outcomes in terms of time, cost and quality due to the co-operative project arrangement.

United States

Rojas and Kell (2008) investigated around 300 school projects in the Northeast of the United States. The actual project cost was higher than the GMP value in 75% of the cases. On the other hand, Bogus et al. (2010) analysed the performance of public water and wastewater facilities in the United States. The study showed that when the cost and time aspects were compared with lump-sum contracts, contracts using cost-plus fee with the GMP contract performed more desirably in terms of cost and time.

Hong Kong

D.W.M. Chan et al. (2011b) investigated an underground railway station modification and extension works project under TCC procurement through interviews. They found that the project saved 5% in cost and 20% in time. They also analysed a prestigious private commercial development procured with a GMP contract. Their findings showed that the project aligned the individual goals of project stakeholders, achieved a cost saving of 15% and finished six days earlier than the schedule.

Key performance indicators in construction

Swan and Kyng (2004) define a key performance indicator (KPI) as a means for indicating the performance of a project or a firm against critical criteria. Cox et al. (2003) also defined KPIs as compilations of data measures for assessing the performance of a construction projects. KPIs enable the evaluation of organisational and project performance in the construction industry (KPI Working Group, 2000). A number of research studies on KPIs within the construction industry are analysed in the construction management literature. Table 10.1 summarises the KPIs from some previous literature from 2000 to 2012 (Chan and Chan, 2012). In addition, Table 10.1 shows the meanings of the KPIs, and the number is limited to thirty for easy reference.

In response to the Egan (1998), the KPI Working Group (2000) collected opinions regarding performance assessment of the United Kingdom construction industry. A total of thirty-eight indicators suggested by the report were divided into six categories to measure the performance of the whole supply chain in a construction project. It provided a flexible framework enabling various stakeholders' companies along the entire supply chain, such as suppliers, subcontractors, main contractors, consultants and employers, to apply individual KPIs to suit their specific needs. Cox et al. (2003) undertook a study of KPIs via a questionnaire survey of project managers

Table 10.1 Summary of the key performance indicators to evaluate the success of construction projects worldwide (Chan and Chan, 2012) (with permission from Sweet & Maxwell)

	KPIs	KPI Working Group (2000)	Nicolini et al (2001)	Cox et al. (2003)	Swan and Kyng (2004)	Cheung et al. (2004)	Menches and Hanna (2006)	Lam et al. (2007)	Kalarachbi and Jones (2008)	Jones and Kalarachbi (2008)	Luu et al. (2008)	Rojas and Kell (2008)	Tennant and Langford (2008)	De Marco et al. (2009)	Toor and Ogunlana (2010)	Chan et al. (2010d)	D.W.W. Chan et al. (2011b)	Haponava and Al-Jibouri (2012)	Total no. of hits for each KPI identified
Time	1 Time for construction	✓	✓	✓	✓	✓	✓	✓	✓	✓	✓			✓	✓	✓	✓		14
	2 Time predictability – design and construction	✓			✓	✓			✓	✓						✓		✓	7
Cost	3 Time to rectify defects	✓		✓															2
	4 Cost for construction	✓	✓	✓	✓	✓	✓	✓	✓	✓	✓			✓		✓			11
	5 Cost exceeding GMP/target cost or not		✓				✓							✓			✓		4
	6 Cost predictability – design and construction	✓			✓	✓						✓	✓			✓			6
	7 Occurrence and magnitude of disputes and conflicts							✓							✓	✓	✓		4
	8 Cost of superstructure								✓	✓									2
	9 Development fee								✓	✓									2
	10 Consultant fee								✓	✓									2
	11 Cost per m²			✓															1
	12 Number of change orders generated	✓				✓	✓												3
Quality	13 Quality	✓			✓	✓		✓									✓	✓	4
	14 Defects (number/severity)	✓	✓	✓	✓	✓			✓	✓							✓		9
	15 Quality issues at end of defect rectification period	✓							✓	✓									3
	16 Quality management system										✓								1
	17 Aesthetics							✓	✓										2

(continued)

Table 10.1 (continued)

Category	KPIs	KPI Working Group (2000)	Nicolini et al (2001)	Cox et al. (2003)	Swan and Kyng (2004)	Cheung et al. (2004)	Menches and Hanna (2006)	Lam et al. (2007)	Kaluarachchi and Jones (2008)	Jones and Kaluarachchi (2008)	Luu et al. (2008)	Rojas and Kell (2008)	Tennant and Langford (2008)	De Marco et al. (2009)	Toor and Ogunlana (2010)	Chan et al. (2010d)	D.W.M. Chan et al. (2011b)	Haponava and Al-Jibouri (2012)	Total no. of hits for each KPI identified
Satisfaction	18 Client's satisfaction	✓			✓				✓	✓	✓		✓					✓	7
	19 Contractor's satisfaction								✓	✓									2
	20 Conformance to stakeholders' expectations								✓	✓					✓	✓	✓	✓	6
Health, safety and environment	21 Safety	✓			✓	✓		✓	✓	✓	✓				✓		✓		9
	22 Reportable accidents		✓			✓									✓	✓			4
	23 Lost time accidents					✓		✓											2
	24 Environmental performance					✓		✓	✓	✓							✓		5
	25 Quantity of waste generated					✓			✓	✓							✓		4
Other	26 Contractor involvement					✓			✓							✓	✓	✓	5
	27 Productivity performance			✓	✓	✓													3
	28 Staff turnover			✓															1
	29 Training days												✓						1
	30 Profit predictability (project)	✓			✓														2
	Total no. of KPIs identified from each publication	12	4	6	9	13	3	7	17	15	5	1	5	2	6	6	11	6	128

and senior construction executives. It was found that these two groups of participants had different views of quality control and on-time KPIs. The survey found that project managers placed more emphasis on the project level, while senior executives tended to focus on the company-wide level.

Swan and Kyng (2004) provided useful guidelines for benchmarking construction projects, stating that if the requirements for a KPI system are put in place, it is essential to decide which to assess. Their study suggested that the number of KPIs should be limited to eight to twelve, otherwise the performance measurement exercise would be challenging and the collection of the necessary data would also be onerous. In addition, the data should be collected reasonably. If it is foreseeable that no action will be taken wherever the KPIs are high, or low, they should not be considered as real 'key' performance indicators. They further commented that most performance measurement systems involve a combination of external benchmarks (e.g. productivity and safety) and internal benchmarks (e.g. time needed for settlement of the final project account). The inclusion of internal benchmarks could help users to compare among their own projects, but not at a national or industry-wide level.

Cheung et al. (2004) established a Web-based construction project performance supervision system to help project managers exercise that role. Eight project performance factors were listed for their performance monitoring system, involving people, time, cost, quality, safety and health, environment, employer's satisfaction and communication. The performance indicators and their relative measurements were generated under each category in the system. Menches and Hanna (2006) carried out a study on quantitative measurement of successful performance from project managers' perspective in the United States. Firstly, they carried out fifty-five interviews to define project success from the project managers' views. Questionnaires were then delivered to find out the variables for inclusion in a performance measurement index. Lam et al. (2007) generated a project success index to benchmark the performance of construction projects under the design-and-build procurement strategy according to four KPIs (time, cost, quality and functionality). Information was gained from forty design-and-build projects in Hong Kong. A project success index curve was then generated. Lam et al. (2007) suggested that construction companies could compare their own project performance levels with counterparts by referring to the corresponding scores along the curve.

Jones and Kaluarachchi (2008) established a multi-dimensional benchmarking model for social house building innovation projects in the United Kingdom. The model showed the performance of social housing provisions by considering the demand and supply aspects of the development stages through the benchmarking model. Luu et al. (2008) suggested a conceptual framework for benchmarking the project management performance from the builder's view in Vietnam. The study identified nine KPIs to assess the builders themselves and their competency. They conducted case studies of three major contractors in order to verify the validity of the model, and claimed that it could be used by other builders with only minor changes.

Tennant and Langford (2008) analysed case studies of three construction companies involving thirteen projects. Their results suggested that the use of performance management systems could benefit construction managers for project appraisal. Chan (2009) used the balanced scorecard approach to study the link between critical success factors and strategic thrusts defined in the Construction Industry Master Plan in Malaysia. Eight critical success factors and seven strategic thrusts were involved in the master plan. They generally covered the four perspectives under the balanced scorecard approach (financial, customer, internal, and learning and growth perspectives, with a strong emphasis on learning and growth.

A case study conducted by De Marco et al. (2009) showed the use of index-based estimate and logistic estimates for both cost estimate at completion and time estimate at completion in an industrial project in Turin, Italy. They found that index-based estimates are reliable for project cost control, while time at completion is better estimated with logistic models. Toor and Ogunlana (2010) carried out a questionnaire survey in Thailand regarding nine KPIs for mega-sized infrastructure projects, and analysed the importance of the KPIs from the perspectives of various stakeholders, including clients, builders and consultants. Their results showed that safety, efficient use of resources, reduced conflicts and disputes and the like become increasingly important. They also suggested that the construction industry has been slowly changing from conventional performance measurement to a combination of both quantitative and qualitative performance measurements on large-scale infrastructure projects. Haponava and Al-Jibouri (2012) generated a generic system for evaluating project performance based on a series of process-based KPIs relating to both process completeness and process quality at the pre-project, design and construction stages.

Performance measurement of TCC under NHS ProCure21+ Framework

The United Kingdom National Health Service implemented the ProCure21+ Framework, in which NEC3 Option C (TCC with Activity Schedule) is adopted, for capital investment construction projects (NHS, 2011). In the procurement framework, a performance management system is employed and covers the following six major areas of KPIs:

1 time certainty;
2 cost certainty;
3 employer's satisfaction (on products);
4 employer's satisfaction (on services);

5   health and safety;
6   defects.

Nevertheless, a large amount of information and data are needed to compute the KPIs identified. For instance, elemental cost breakdown is required, and the evaluation of the Achieving Excellence Design Evaluation Toolkit should be completed. Moreover, the United States has well-developed TCC and GMP procurement methods, but the use of TCC and GMP schemes is still in its infancy in Hong Kong. Thus, by comparing TCC and GMP strategies in Hong Kong with those from the ProCure21+ Framework, the similarities or differences in the performance measurement systems of the two jurisdictions could be observed. As the performance measurement framework of ProCure21+ is employer-driven and a large amount of information and data have to be entered into the assessment tool (e.g. cost data should be entered into the elemental cost analysis of the framework), it could be made more user-friendly and effective to establish an overall performance index to measure the performance levels of projects procured with TCC and GMP, instead of comparing the KPIs individually. Drawing an analogy, the performance of a primary school student is evaluated based on his or her overall score/position in class (overall performance index). This method would be more direct and holistic than comparing the number of distinctions gained in individual subjects (individual KPIs).

Round 1: identifying the most important KPIs

Based on a comprehensive review of the literature about performance measurement in construction projects, a questionnaire was designed to solicit the views of an expert panel with Delphi survey. For details of the procedures of the questionnaire survey, please refer to Chan and Chan (2012). Table 10.2 indicates the relative importance of all the key

Table 10.2 Results of Round 1 Delphi survey (Chan and Chan, 2012) (with permission from Sweet & Maxwell)

KPIs for TCC and GMP construction projects	Total frequency	%
Mutual trust between project partners	15	93.75
Time performance	14	87.50
Magnitude of disputes and conflicts	11	68.75
Final out-turn cost exceeding the final contract target cost or guaranteed maximum price value or not	11	68.75
Client's satisfaction with quality of completed work	11	68.75
Contractor's feedback on client's decision making process	9	56.25

(continued)

Table 10.2 (continued)

KPIs for TCC and GMP construction projects	Total frequency	%
Time required for the settlement of final project account	10	62.50
Contractor's involvement in project design	8	50.00
Design quality	8	50.00
Time needed from the commencement of project design up to contract award	6	37.50
Percentage of contractor's alternative design proposals approved by consultants at first attempt	6	37.50
Safety performance	5	31.25
Contractor's satisfaction on TCC and GMP contractual arrangement	6	37.50
Environmental friendliness	2	12.50
Cost per m² of construction floor area, including foundations	1	6.25
Form of contract to be used	1	6.25
Contractor's ability to perform cost management	1	6.25
Appropriateness of risk allocation	1	6.25
Time allowed for pre-construction preparation works	1	6.25
Contractor's claim-consciousness	1	6.25
Amount of works that the tenderer has in hand at the final stage of tendering	1	6.25

performance indicators. A total of fifteen KPIs were listed on the survey form, and the remaining six (shown in italics) were suggested by the expert panel. The table also reveals the frequencies of hits for the corresponding indicators.

Round 2: refining the selected KPIs

Subsequent to Round 1, questionnaires were sent by mail to the participants on the expert panel in May 2011. In Round 2, the findings of Round 1 were consolidated and provided to the experts, and they were invited to reconsider whether they would like to make some changes to their initial choices as a result. Fourteen members responded, and two experts withdrew from the study.

As shown in Table 10.3, seven KPIs were identified with more than 50% frequency by the Delphi panel of participants (shown in bold). Therefore, a total of seven most significant KPIs was selected specifically for gauging the performance of TCC and GMP strategies in Hong Kong, in descending order: (1) 'mutual trust between project partners', (2) 'time performance', (3) 'final out-turn cost exceeding the final contract target cost, or GMP value, or not', (4) 'magnitude of disputes and conflicts', (5) 'client's satisfaction with quality of completed work', (6) 'time required for the settlement of final project account' and (7) 'contractor's involvement in project design'. The remaining four KPIs mainly emphasise the working relationship between the client and the builder.

Round 3: establishing individual weightings for the seven most important KPIs

In Round 3 of the Delphi questionnaire, panel members were requested to assign a level of significance (rating) to the top seven identified KPIs by means of a five-point Likert scale (1 = least important, 2 = slightly important, 3 = important, 4 = very important and 5 = most important) to gauge the performance of TCC and GMP projects. In July 2011, following some email reminders, all fourteen panel experts submitted their completed questionnaires.

Table 10.3 Results of Round 2 Delphi survey (Chan and Chan, 2012) (with permission from Sweet & Maxwell)

KPIs for TCC and GMP construction projects	Total frequency	%
Mutual trust between project partners	14	100.00
Time performance	12	85.71
Final out-turn cost exceeding the final contract target cost or guaranteed maximum price value or not	11	78.57
Magnitude of disputes and conflicts	10	71.43
Client's satisfaction with quality of completed work	10	71.43
Time required for the settlement of final project account	10	71.43
Contractor's involvement in project design	10	71.43
Contractor's feedback on client's decision making process	6	42.86
Design quality	6	42.86
Time needed from the commencement of project design up to contract award	6	42.86
Percentage of contractor's alternative design proposals approved by consultants at first attempt	4	28.57
Safety performance	4	28.57
Contractor's satisfaction with TCC and GMP contractual arrangement	3	21.43
Contractor's ability to perform cost management	2	14.29
Appropriateness of risk allocation	2	14.29
Contractor's claim-consciousness	2	14.29
Form of contract to be used	1	7.14
Time allowed for pre-construction preparation works	0	0.00
Amount of works that the tenderer has in hand at the final stage of tendering	0	0.00
Environmental friendliness	0	0.00
Cost per m² of construction floor area, including foundations	0	0.00

Notes: Additional KPIs suggested by the expert panel are in *italics*. KPIs with a percentage of 50% or higher are shown in **bold**.

A statistical analysis was conducted based on the fourteen survey forms in order to establish the mean ratings of the seven most essential KPIs. Therefore, a primary basket of the most significant KPIs and their respective weightings were established according to the mean ratings scored by the panel members. Each of the seven KPIs was calculated based on a measurement scale of score of 1–5, where 1 denoted least important and 5 denoted most important to gauge the success of TCC and GMP projects. The weighting of each KPI was computed as its individual mean rating divided by the total mean ratings of all the KPIs, calculated using the equation below. This method has been used by several researchers before, such as Chow (2005), Yeung et al. (2007), Yeung et al. (2009), Eom and Paek (2009) and Tennant and Langford (2008).

$$W_{KPIa} = \frac{M_{KPIa}}{\sum\limits_{g} M_{KPIg}} \text{ for } a = 1$$

where:

W_{KPIa} represents the weighting of a particular top seven KPI in Round 3;
M_{KPIa} represents the mean rating of a particular top seven KPI in Round 3;
$\sum\limits_{g} M_{KPIg}$ represents the summation of the mean ratings of all the top seven KPIs in Round 3.

Table 10.4 summarises the seven most significant KPIs and their corresponding weightings.

The seven KPIs were: (1) 'mutual trust between project partners', with a weighting of 0.180, (2) 'final out-turn cost exceeding the final contract

Table 10.4 Results of Round 3 Delphi survey (Chan and Chan, 2012) (with permission from Sweet & Maxwell)

KPIs for TCC and GMP construction projects	Rank	Mean rating	Corresponding weighting
Mutual trust between project partners	1	4.71	0.180
Final out-turn cost exceeding the final contract target cost or guaranteed maximum price value or not	2	4.14	0.158
Time performance	3	4.07	0.155
Magnitude of disputes and conflicts	4	3.50	0.134
Client's satisfaction with quality of completed work	4	3.50	0.134
Contractor's involvement in project design	6	3.21	0.123
Time required for the settlement of final project account	7	3.07	0.117
Number of respondents	**14**		

Note: Mean rating: 1 = least important; 5 = most important.

target cost, or GMP value, or not', with a weighting of 0.158, (3) 'time performance', with a weighting of 0.155, (4) 'magnitude of disputes and conflicts', with a weighting of 0.134, (5) 'client's satisfaction with quality of completed work', also with a weighting of 0.134, (6) 'contractor's involvement in project design', with a weighting of 0.123, and (7) 'time required for the settlement of final project account', with a weighting of 0.117. A composite performance measurement index (PMI) for TCC and GMP construction projects in Hong Kong was hence calculated based on the following equation:

Performance Measurement Index (PMI)

$= 0.180 \times$ Mutual trust between project partners

$+ 0.158 \times$ Final out-turn cost exceeding the final contract target cost, or guaranteed maximum price value, or not

$+ 0.155 \times$ Time performance

$+ 0.134 \times$ Magnitude of disputes and conflicts

$+ 0.134 \times$ Client's satisfaction with quality of completed work

$+ 0.123 \times$ Contractor's involvement in project design

$+ 0.117 \times$ Time required for the settlement of final project account

The PMI comprised the seven weighted KPIs identified in Round 2 of the Delphi survey, and their weightings were computed as the individual mean scores divided by the total mean scores. It was assumed that the PMI was linear and additive. The units of measurement of the seven KPIs are different, so there is not likely to be a multiplier effect among them (Hughes et al., 2011; KPI Working Group, 2000). It would be simpler and easier to apply this linear model equation in practice to evaluate the performance standards of TCC and GMP strategies in Hong Kong construction projects.

Round 4: re-evaluating the weighted KPIs in Round 3

The Round 4 results were based on the consolidation of Round 3 results by the fourteen respondents. They were given the mean ratings of the fourteen members for each KPI and their own individual ratings suggested in Round 3. Each member was then invited to decide whether to modify the initial choices after referring to the mean rating of all fourteen participants.

As Table 10.5 shows, there was no change in ranking for most KPIs, but the rankings of 'time required for the settlement of final project account' and 'contractor's involvement in project design' were interchanged compared with Rounds 3 and 4. Moreover, their respective weightings were similar to those in Round 3. Therefore, PMI could be re-computed as in the following revised model equation:

Performance Measurement Index (PMI)

= 0.176 × Mutual trust between project partners

+ 0.163 × Final out-turn cost exceeding the final contract target cost, or GMP value, or not

+ 0.158 × Time performance

+ 0.136 × Magnitude of disputes and conflicts

+ 0.131 × Employer's satisfaction with quality of completed work

+ 0.120 × Time required for the settlement of final project account

+ 0.115 × Contractor's involvement in project design

Findings of the Delphi survey

After analysing four rounds of the Delphi survey, the top seven weighted KPIs of TCC and GMP contracts focused broadly on project success, relationships and people. Conventionally, project success is evaluated based on time, cost and quality (Chan and Kumaraswamy, 2002). The results were in line with this factor, as time, cost and quality performance were ranked third, second and fifth respectively. The findings also emphasise the importance of relationships and people. It is beyond doubt that the top four weighted KPIs in Table 10.5 are: (1) 'mutual trust between project partners', (2) 'magnitude of disputes and conflicts', (3) 'time required for the

Table 10.5 Results of Round 4 Delphi survey (Chan and Chan, 2012) (with permission from Sweet & Maxwell)

KPIs for TCC and GMP construction projects	Rank	Mean rating	Corresponding weighting
Mutual trust between project partners	1	4.71	0.176
Final out-turn cost exceeding the final contract target cost or guaranteed maximum price value or not	2	4.36	0.163
Time performance	3	4.21	0.158
Magnitude of disputes and conflicts	4	3.64	0.136
Client's satisfaction with quality of completed work	5	3.50	0.131
Time required for the settlement of final project account	6	3.21	0.120
Contractor's involvement in project design	7	3.07	0.115
Number of respondents	**14**		

Note: Mean rating: 1 = least important; 5 = most important.

settlement of final project account' and (4) 'contractor's involvement in project design'. They are the major objectives of stakeholders administrating TCC and GMP projects. These findings echo previous studies regarding KPIs for TCC and GMP construction projects (Cox et al., 2003; Nicolini et al., 2001; American Institute of Architects, 2001). The top seven KPIs will be briefly introduced below.

Mutual trust between project partners

Wong and Cheung (2005) considered that mutual trust is important to achieve an efficient partnering strategy. Black et al. (2000) conducted partnering studies, and summarised that mutual trust among stakeholders was essential to the success of the partnering strategy. Partnering is always applied with TCC and GMP contracts in Hong Kong (Chan et al., 2010d; American Institute of Architects, 2001; Anvuur and Kumaraswamy, 2010). Yeung et al. (2009) carried out similar research regarding the evaluation of the success of partnering projects in Hong Kong by applying a Delphi survey study, and also found 'mutual trust and respect' to be one of the major KPIs for partnering projects.

Time performance

Time performance is perceived as a common KPI around the globe (KPI Working Group, 2000; Jones and Kaluarachchi, 2008). Lam et al. (2007) considered time as one of the KPIs for design-and-build construction projects in Hong Kong. In a TCC underground railway extension project, time performance was also an important KPI to evaluate its success (Hughes et al., 2011) and that of a private office building project (Cox et al., 2003) in Hong Kong. Frampton (2003) revealed that projects procured with TCC and GMP could enable construction activities to begin before completing the design phase. It would be illuminating to observe whether TCC and GMP strategies can perform well in terms of time certainty.

Final out-turn cost exceeding the final contract target cost, or GMP value, or not

The major characteristic of TCC and GMP is to motivate contractors to save costs by aligning the individual interests of the client and contractor (Rose and Manley, 2010). Pursuant to the basic principle under TCC and GMP approaches, both contractual agreements offer motivation to service providers to minimise costs during project delivery by linking the benefits of clients with those of service providers. It is reasonable that the cost performance (i.e. whether actual cost exceeding the final contract target cost, or GMP value, or not) of TCC and GMP projects becomes an important KPI (Chan et al., 2010a; American Institute of Architects, 2001; De Marco et al., 2009).

Magnitude of disputes and conflicts

This result is in line with the findings of Lam et al. (2007) and Toor and Ogunlana (2010). TCC and GMP strategies are usually applied together with partnering method (Mahesh, 2009). Partnering can be an effective strategy to improve the contractual relationship in order to build a cohesive, integrated project team with the same targets and well-defined procedures for dispute resolution using a timely and efficient approach (Bench et al., 2005). It will be important to gauge whether TCC and GMP methods can effectively reduce disputes or confrontations among project stakeholders. The magnitude of disputes and conflicts could be related to the relationship between the employer and builder, depending on the gain-share/pain-share mechanism under TCC and GMP procurements.

Employer's satisfaction with quality of completed work

Quality means conformity to contract specifications and the employer's satisfaction with the built facilities. This factor always ranks among the top priorities of construction projects (Soetanto et al., 2001). It is not surprising that the quality of completed work was regarded as a KPI, as the same finding has been reported in much of the literature (Cheung et al., 2004; Jones and Kaluarachchi, 2008).

Time required for the settlement of final project account

A research study by Yiu et al. (2005) evaluated the performance of consultants in the Hong Kong construction industry based on four stages:

1 design/planning;
2 tender process;
3 construction;
4 final account.

The final account stage showed that settlement of the final account helped to achieve the success of a construction project. Chan et al. (2010d) revealed that TCC and GMP strategies allow early settlement of the final project account. This KPI, 'time required for the settlement of final project account', could be used to assess whether TCC and GMP projects could materialise this benefit.

Contractor's involvement in project design

The importance of involving construction expertise at the design stage has been supported by the construction industry (Song et al., 2009). Mosey (2009) echoed that the design process should not only involve the design consultants, but also contractors and specialist suppliers, to generate a thorough and functional design. It is especially essential for projects procured with TCC and GMP methods, because the contractor can contribute earlier, for instance during design development (Mahesh, 2009). The builder's participation in the design stage could influence the success of a proposed project in terms of time, cost and quality.

Chapter summary

A plethora of literature regarding the performance measurement of construction projects has developed over recent decades. This chapter has established a holistic framework for gauging overall performance, and identified the seven most significant KPIs for TCC and GMP projects in Hong Kong as follows:

1 'mutual trust between project partners';
2 'time performance';
3 'final out-turn cost exceeding the final contract target cost, or guaranteed maximum price value, or not';
4 'magnitude of disputes and conflicts';
5 'client's satisfaction with quality of completed work';
6 'time required for the settlement of final project account';
7 'contractor's involvement in project design'.

Moreover, the established performance measurement tool could help compute a composite performance measurement index to provide a single measure of project performance. A benchmark for the assessment of the overall performance levels of TCC and GMP projects in Hong Kong was also developed in this study. Industrial practitioners can simply input some essential values of individual project performance measures, then compare the performance levels of different TCC and GMP projects within the company or the whole construction industry, or between companies, to observe where their TCC and GMP projects stand in relative terms. After that, the different performance levels of the projects can be evaluated and compared objectively on the same basis for benchmarking at project completion, and can be managed in the delivery process of the project.

Senior management and project managers could utilise the PMI to measure, manage, assess and improve performance levels to strive for construction excellence with optimal performance. In addition, the developed performance measurement tool could enrich existing knowledge regarding the KPIs for TCC and GMP strategies.

Part III

Case studies in target cost contracts

Part III

Case studies in larger
cost centres

11 Case studies in target cost contracts

TCC: London 2012 Olympic Park developments, England

Background

The UK Olympic Delivery Authority (ODA) was founded to design and deliver the London 2012 Olympic Park project for hosting the London 2012 Games (Davies and Mackenzie, 2014). Although the economy of the UK was declining at that time and there was severe public scrutiny and interest, the project was still completed within budget and before the forecast complete date. In 2006, the ODA decided to apply the NEC3 suite of contracts to most of its programme and utilise the NEC's spirit to maintain a healthy relationship between itself and CLM, its delivery partner for the projects (Davies and Mackenzie, 2014).

NEC3 contracts provide a high degree of flexibility, as well as methods to change control, enabling genuine, distinctive co-operation among various stakeholders in the construction industry. It also facilitated the project delivery via a transparent and governable mechanism.

As ODA started its business in 2006, it encountered substantial hindrances. A site of around 2.5km² needed to be redeveloped (Bryant, 2012).

The final cost of this project was £6.8 billion, which was within the original budget (£8.1 billion) allocated to complete the London 2012 Olympic Park (Davies and Mackenzie, 2014). In addition, the project was completed on time in July 2011, which allowed thirteen months for testing the venue before the games commenced on 27 July 2012 (Davies and Mackenzie, 2014).

Motives for using TCC

According to NEC Users' Group (Bryant, 2012), the motives for adopting TCC by the ODA were as follows:

- a common approach in the current construction market, with a proven track record in the supply chain;
- a motivation to collaborate;

- a comprehensive system of contractual procedures, supporting various project types;
- high transparency of risk management;
- a vigorous change control mechanism providing transparency over the time and cost impacts of every change event;
- an equitable risk sharing mechanism covering potential variations to the Olympic Park and the fluctuating economic climate via adopting NEC3 Engineering and Construction Contract (ECC) Option C (TCC with activity schedule);
- encouraging best practice in the construction industry and providing more flexible solutions for contracting out different degrees of risk;
- encompassing and integrating the stages to ensure corporate governance, such as the obligation to manage public funds adequately and openly when lodging contract changes which fall within the compensation events;
- providing transparency in change control and a contractual route to mitigate risks and realise opportunities at contract level in the early warning notice procedure;
- encouraging a co-operative and active atmosphere in contract administration;
- enabling the ODA to develop a practical platform where it could adopt a consistent contracting structure throughout the supply chain via using the standardised suite of London 2012;
- allowing selection of the best suitable risk profile, balancing time, cost and quality against the nature of the project and its design development stage during procurement.

Perceived benefits of NEC3

The NEC Users' Group (Bryant, 2012) identified the following benefits when using NEC3.

Mutual trust and co-operation

ECC Clause 10 identifies and restricts the behaviour of the stakeholders under the contract, in that every stakeholder has to work collaboratively in 'a spirit of mutual trust and co-operation'. Hence, NEC3 promotes a co-operative and proactive working relationship for contract administration.

Good corporate governance

The NEC integrates a set of procedures to ensure corporate governance. For instance, there is an obligation to manage public funds transparently and properly when contract change is initiated, which will be tackled via the compensation event process.

Encouraging discipline

As the NEC is well known as being onerous to administer, it requires discipline for the timely operation of contract mechanisms, such as the limitation of time for reply within the contract administration stage to enhance contemporaneous resolution of matters.

Early warning

The contract administration procedures and the contract management software tools enable speeding up the progress of compensation events, enabling early warning notices and project manager assessments and decisions to be supervised efficiently. Moreover, active supervision allows prompt intervention, where necessary, to avoid undue delay in the resolution of issues. In addition, the early warning notice can help control changes and provide a contractual method to mitigate risks and realise opportunities at contract level.

Standardisation of documentation

The standardisation of London 2012 contracts helped to ensure a consistent contract form throughout the supply chain. The supply chain became more familiar with NEC3, so it could enhance efficiency as the programme progressed. NEC3 enabled the selection of the most suitable risk profile to balance time, cost and quality using various contract and subcontract options associated with core clauses, secondary option clauses and Z clauses (contract amendments) to suit the works needed.

Contract incentives

NEC3 contracts involve incentive mechanisms through sharing gain and pain to accomplish the tasks. The stated objectives in the ODA's public procurement documents focus on working in partnership with industry to construct sustainable and world-class facilities and achieve value for money. By adopting TCC with activity schedule, the ODA could determine the incentives flexibly by setting completion bonuses and imposing key performance indicators for the stated criteria. This incentive mechanism encouraged the supply chain to utilise potential efficiencies in construction and design. Furthermore, establishing bonus milestones emphasised effort on completion at the stated points in the programme. There were other key performance indicators to evaluate the issues (e.g. sustainability, health and safety, and employment requirements).

As changes to the design were anticipated, a level of flexibility was needed to cope with them. Moreover, the pain/gain mechanism could encourage the Tier 1 contractor to comply with the ODA's objectives and management of risks.

Major difficulties in implementing NEC3

The NEC Users' Group (Bryant, 2012) indicated that NEC3 contracts have long been regarded as onerous to administrate. Contractual change of time and cost, risk mitigation and opportunity realisation need to be proactively and frequently monitored in such a high-profile public project as London 2012, regardless of contract form.

Lessons learned

Some key lessons were learned from this landmark project, according to the NEC Users' Group (Bryant, 2012).

Prepare more resources

TCC was generally perceived as one of the best forms of contract to tackle the extent of change instructed. Nevertheless, TCC required a high degree of trust, confidence, administrative effort as well as understanding.

Acquire sufficient project information

It is crucial to acquire the following information in sufficient detail:

- interfaces with others in and around the construction site;
- completion requirements of sectional completion;
- handover process and/or requirements, in order to ensure that the latest requirements are considered, involving a clear statement showing the party accountable for documentation;
- a clear description of roles and accountabilities in design, planning, licenses and approvals, and utilities;
- clear statements regarding the availability of items, e.g. free-issue materials;
- unequivocal descriptions of what requirements the main contractor should pay attention to, rather than the supply chain.

Ensure tenderers submit an accepted programme at the outset

A realistic programme of work is a key element to achieve a successful TCC contract. The programme is imperative for efficient project management, and has to be effectively supported by project control teams. It is crucial that the Tier 1 and Tier 2 planners genuinely comprehend the project control approaches at the beginning of the contract, involving the concepts of 'planned value' and 'earned value'.

Establish good communication and commercial behaviour at outset

TCC encourages good commercial behaviour, which can be observed in the London 2012 project. The working environment is usually collaborative, with sharing of ideas, procurement initiatives and working practices. CLM (ODA's delivery partner) and the Tier 1 teams co-operated to improve relationships and commitments for delivery. This would be further improved if the Tier 1 programme planners could be co-located.

Ensure tenderers provide detailed supporting information with bids

Apart from the standardised pricing documents, it is suggested that tenderers should submit supplementary information in the form of a 'bid book' with a detailed breakdown of the activity schedule. It was a formal requirement for the contract for the South Park, which substantially decreased the number of clarification questions and alleviated evaluations.

Maintain communication and commitment throughout supply chain

Due to the technical requirements, it was not easy to align the subcontractors with the administrative programme and project management requirements. In addition, time is needed to allow the builders and suppliers to get up to speed on information technology, site access, security, insurances and best practice on large-scale projects. Therefore, training was provided to the contractor and supply chains in the Primavera P6 Enterprise Project Portfolio Management software. All stakeholders believed that the training raised their level of expertise and productivity.

TCC: Mass Transit Railway Corporation Tsim Sha Tsui Metro Station Modification Works Project, Hong Kong

Background

The Tsim Sha Tsui Underground Railway Station Modification and Extension Works project was the earliest TCC in Hong Kong that was entirely 'open-book'. It attempted to innovate and make value engineering a priority via the gain-share/pain-share mechanism of the TCC process. The project connected the pedestrian subway links of the new Kowloon–Canton Railway Corporation (KCRC) East Tsim Sha Tsui Station to the existing Mass Transit Railway Corporation (MTRC) Tsim Sha Tsui Station at the south end, and improved access and egress at the north end (the KCRC merged with the MTRC in December 2007, and the HKSAR Government maintains a majority stake in the MTRC). It also created a single-level extension to one end of the existing underground structure. According to Chan et al. (2010d), the main objectives of the works were:

1 to integrate the Tsim Sha Tsui underground network for daily travellers;
2 to alleviate congestion and increase station accessibility to cope with surges in passengers and new commercial developments in the surroundings;

3   to offer a more desirable travelling environment for passengers;
4   to provide convenient access for passengers with special needs by introducing passenger lifts.

This project was built under Nathan Road, a vital trunk road in one of the
bustling regions, within a cut and cover cofferdam. Other station modification projects aimed to alter the existing station structure substantially and
maintain passenger flows at all times. In a bustling urban district, deep excavation and pedestrian tunnel construction were included in the project. The
excavation for the tunnel was about 1.5m over the crown of a functioning
underground railway tunnel, as well as the temporary retaining structures at
a similar distance from the side of the subways. A high level of construction
risk thus resulted. Risk management and mitigation became an exceptionally
crucial factor in the success of this construction project. The client organisation and project manager was MTRC. The project team involved a Japanese
main contractor, an electrical and mechanical engineering consultant, and
various specialist subcontractors, such as for instrumentation, cladding,
steelwork, ceilings, etc.

Although this project was risky and the schedule was extremely tight, it
was completed within cost and time, which were initially set at HK$300
million and thirty-six months. The target cost rose to HK$312.5 million
due to some variations. The final cost was HK$297.7 million, generating
HK$14.8 million (approximately 5% of cost savings). The project was com-
pleted seven months earlier than the contract completion date. Figure 11.1
shows the time and cost profiles of this project. This case study will effectively

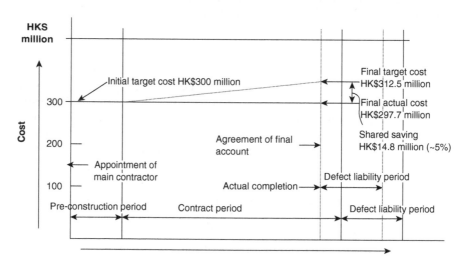

Figure 11.1 The cost and time profiles of the Tsim Sha Tsui Underground Railway
    Station Modification and Extension Works (Avery, 2006).

justify the adoption of alternative integrated contracting strategies to align the suitable project team to the risk profile (Avery, 2006).

Motives for using TCC

Chan et al. (2010d) investigated the following motives in the Tsim Sha Tsui Metro Station Modification Works project using an interview survey.

Providing financial incentives

The Tseung Kwan O Underground Railway Extension project showed that that the use of an incentivisation agreement (IA) can bring benefits to overall project performance. An IA is similar to TCC in principle, where the employer and the builder mutually agree at the outset date that all remaining works from this agreed date onward will be measured with an estimated cost for their risks using the gain-share/pain-share approach. An IA can help the contractor to work efficiently and achieve cost savings. The MTRC believed that using incentive arrangements in the Tsim Sha Tsui Underground Railway Station Modification and Extension Works project was sensible. The mechanism of a fully open-book TCC approach adopting the gain-share/pain-share mechanism was established, with the objective of achieving excellent project outcomes. The employer also intended to use this project as the benchmark model for future target cost-based construction projects, in particular large-scale, technically difficult contracts such as the West Island Underground Railway Line. In addition, the TCC procurement strategy could offer financial incentives to the builder to contribute and save costs by presenting innovative ideas. As this project included substantial uncertainties and a high risk profile, utilising the traditional fixed-price lump-sum contract would likely lead to claims and poor working relationships among the project stakeholders.

Greater time and cost certainty

The adoption of the TCC strategy via the gain-share/pain-share mechanism enabled the achievement of higher certainty on time, cost and quality for the employer and encouraged the contractor to place more emphasis on the management and mitigation of risks.

Improving working relationships

It aimed to significantly improve the working relationships and bring in a more collaborative approach to conflict resolution. The employer wanted to align the overall project parties' objectives by offering the most suitable overall solutions without compromising the safety and operation of the railway, striking a practical balance between the programme and total project cost (Dunn and Jones, 2004). This alternative integrated procurement strategy was likely to minimise contractual claims from contractors.

Perceived benefits of TCC

Strict control over tendering process

It was found that TCC required more strict control over the tendering process, subcontract procurement, risk management, contract administration, and higher transparency for financial control and quality of information needed for early financial planning. These substantially contributed to the outstanding project performance of the Tsim Sha Tsui Underground Railway Station Modification Works.

An appropriate performance-based remuneration was applied in this case. The financial interests of employer and contractor were linked together, so they could work in a more collaborative approach (Wong, 2006).

Aligning individual objectives

The gain-share/pain-share mechanism under TCC aligns the individual objectives of the contracting parties (employers and contractors) to the overall objectives of projects and creates harmonious working relationships within integrated project teams. Due to the agreements under TCC and partnering initiatives, the employer and the builder need to manage their works together and share any consequent profits and losses. The project stakeholders opined that they had more opportunities to express their opinions and concerns openly and freely under the TCC procurement method. The incentivisation agreement established a more proactive, co-operative working relationship between the client and contractor, and reinforced the cultural shift away from conventional adversarial methods of contracting (Ting, 2006).

Incentives to work

TCC emphasises incentivising the contractor to work effectively and achieve cost savings in order to gain better value for money throughout the project development (Boukendour and Bah (2001). According to Lam (2002), contractors can bring in their expertise in the project design of construction materials and methods to maximise the buildability of the project. Moreover, compared with conventional contracts, TCC offers a fairer risk allocation among the contracting parties. Early involvement of the contractor is required to help identify and allocate the risks, as advocated by Dunn and Jones (2004). In addition, the adoption of an open-book accounting regime enables quantification of the risks and prevents the risks from negatively affecting the contractor's cash flow (Wong, 2006).

Major difficulties in implementing TCC

Lack of confidence in TCC

After adopting TCC for a project, the rationale behind it needed to be explained to the board of directors of MTRC and to the HKSAR Government, which was the main stakeholder. Obtaining endorsement from the board was demanding, according to Avery (2006). To deal with a high-risk profile project such as this, the common solution in Hong Kong would be a design-and-build lump-sum contract with the entire risks being transferred to the contractor. It was difficult to implement a fully cost-reimbursable TCC with a gain-share/pain-share mechanism as it was a novel approach in Hong Kong at that time. The management team accepted the TCC approach since it was confident that the issue of cost reimbursement would be closely controlled.

Absence of standard form of contract

During the tender stage, there was a lack of a standard or appropriate form of TCC contract within the MTRC internal standard contract agreements. Moreover, the parties lacked experience in implementing the new form of contract (ECC). Therefore, the existing MTRC standard contract was changed. Sadler (2004) found that working in this novel way might not be attractive to the contracting parties, since they might find it uncomfortable and difficult to change the conventional approach. According to Gander and Hemsley (1997), the lack of a standard form of target cost contract leads to a greater possibility of drafting errors and misunderstanding of liabilities among project stakeholders.

Disputes in determining types of variations

Disputes arose at the construction stage as to whether the architects' or engineers' instructions constituted target cost variations or were deemed to be design development, because of ambiguous scope of work (Chan et al., 2007b). Adjudication meetings, including representatives of the employer, engineer and main contractor, were established with the partnering facilitator and relevant stakeholders to resolve contentious issues and intractable disputes.

Critical success factors/lessons learned

Maintain good relationships among contracting parties

Interviewees (Chan et al., 2007b) expressed a consistent view that the overall project success was attributed to the desirable working relationships between the contracting parties and the TCC, which helped establish mutual goals and common interests, as well as an open-book accounting approach. A partnering consultant was selected to help facilitate the team building, improve communication among the project team members and to manage project progress regularly. Building integrated and devoted teams helped to achieve smooth project delivery as well as a fair risk sharing mechanism. The use of a 'shared' site office for the whole project team can further catalyse communication and integration between the stakeholders in a teamwork culture. Tay et al. (2000) pointed out that in order to achieve success under TCC, there should be a genuine willingness to achieve collaboration or show a partnering spirit amongst the co-operating parties.

Appropriate selection of project team

Appropriate recruitment of the project team is thus vital in enhancing mutual trust, productive communication, effective collaboration and efficient conflict resolution (Chan et al., 2004). Under TCC in this case, the employer participated in subcontractor selection. Moreover, a similar target cost contractual approach was adopted in choosing the mechanical and electrical subcontractors. A proactive contractor with strong leadership skills was important to tackle any unanticipated problems and possible disputes (Avery, 2006).

Transparency of the project development process

Another essential ingredient for TCC was transparency throughout the entire project development process. The project parties decided from the beginning that there was to be one set of records from the initial project stage for the contracting parties. Mutual trust and a close working relationship were thus of paramount importance in achieving the open-book accounting regime. Moreover, due to this unique mechanism of TCC based on joint determination and agreement between the employer and the builder on risk allocation, the employer gained realistic target cost estimates, including sufficient risk contingencies under the pain-share/gain-share arrangement.

Provide sufficient incentives

Sadler (2004) suggested that as well as allocating the risks equitably, employers have to gauge the appropriate combination of fee and share, and ensure that there are sufficient incentives to motivate contractors to contribute their innovative ideas in project delivery. It was preferable if the contractor's share was equal to or more than 50%, as suggested by Perry and Barnes (2000). Tang and Lam (2003) recommended a different proportion of sharing to suit various extents of cost savings between the employer and contractor under TCC, as shown in Table 11.1. As suggested by Broome and Perry (2002), a suitable contractual arrangement should focus on aligning

Table 11.1 Suggested share saving percentage apportionment for target cost-type contracts (Tang and Lam, 2003) (with permission from *Hong Kong Engineer*)

Scenario	Client's share	Contractor's share
Final out-turn cost < final target cost		
(a) Saving < 5%	67%	33%
(b) Saving = 5–10%	50%	50%
(c) Saving > 10%	33%	67%

Table 11.2 Summary of the primary attributes associated with the TCC scheme for the Tsim Sha Tsui Underground Railway Station Modification and Extension Works (Chan et al., 2010d) (with permission from Emerald Group Publishing Ltd)

Project nature	Underground railway station modification and extension works involving the connection of the pedestrian subway links in Tsim Sha Tsui, Kowloon, Hong Kong
Contracting approach	TCC approach using two-stage tendering process
Gain-share arrangement	Client:contractor = 50:50
Pain-share arrangement	Client:contractor = 50:50
Underlying motives	• To achieve excellent project performance • To generate financial incentives for the contractor to contribute and save costs by offering innovative ideas • To improve working relationship through partnering spirit • To introduce a more co-operative approach to conflict resolution and minimise claims • To align individual objectives of various contracting parties with the overall project objectives
Key benefits	• Provision of financial incentives for contractor to work efficiently and to achieve cost savings • More rigorous control over tendering process, subcontract procurement, risk management and contract administration • Greater transparency for financial control and higher quality of information exchange • Harmonious working relationship within the project team via partnering arrangement • Development of common overall project goals among various project stakeholders • Enhanced buildability of project design • More equitable risk apportionment between client and contractor
Major difficulties	• Unfamiliarity with or misunderstanding of TCC concepts and practices by senior management • Lack of a suitable form of contract for TCC in the local context • Dispute (claim) occurrence due to unclear scope of work in client's project brief

(continued)

Table 11.2 (continued)

Critical success factors	• Good working relationship and right selection of project team • Shared objectives with common interests • Open-book accounting arrangement in support of tender pricing by contractor • Strong leadership and proactive contractor • Transparency of the entire project development process

the interests of the stakeholders in order to achieve the project goals and take into account the constraints, risks, strengths and weaknesses of the various stakeholders. The contract and incentive structures should match the project objectives and circumstances (Bower et al., 2002). Table 11.2 shows the major interview findings based on the above qualitative analysis of this project (Chan et al., 2010d).

TCC: Open Nullah Improvement Project at Fuk Man Road in Sai Kung, Hong Kong

Background

The employer, project manager, supervisor, main contractor, and NEC adviser and partnering workshop facilitator for the Improvement of Fuk Man Road Nullah Project in Sai Kung were the Drainage Services Department of the HKSAR Government, Chief Engineer (Drainage Projects) from the Drainage Services Department, Black & Veatch, Chun Wo Construction & Engineering Company Ltd, and JCP Consultancy International Ltd respectively. This project was the first HKSAR Government pilot project to promote the New Engineering Contract in order to encourage a partnering and co-operative working relationship between client and contractor. The form of the main contract selected was ECC Option C: Target Cost with Activity Schedule. Figure 11.2 illustrates the scope of works for this project (Chan et al., 2014). The scope of works included the decking of an existing open nullah, which was around 180m long and 12m wide, at Fuk Man Road in Sai Kung, the development of a 4,000m² urban recreational area above the top and the enhancement of a neighbouring roundabout improvement works. The project was initiated in August 2009 and completed twenty-four weeks before schedule with a final project cost of approximately HK$72.9 million.

Key performance indicators

Five key performance indicators were mutually agreed and established to assess these aims throughout the course of construction.

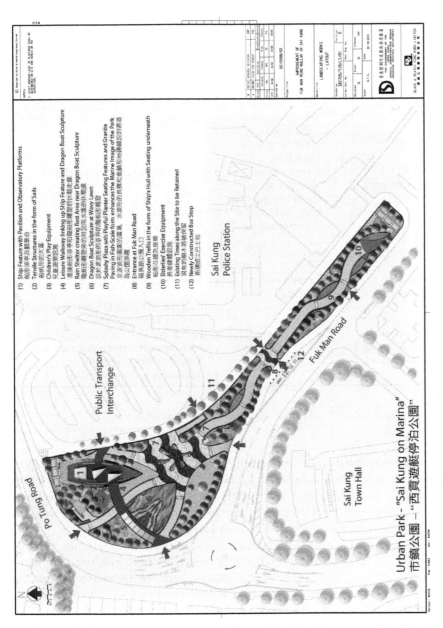

(1) Ship Feature with Pavilion and Observatory Platforms
    船形涼亭及觀景台
(2) Tensile Structures in the form of Sails
    帆桅形的天幕
(3) Children's Play Equipment
    兒童遊樂設施
(4) Leisure Walkway linking up Ship Feature and Dragon Boat Sculpture
    連達船形涼亭和龍舟雕塑的休閒走廊
(5) Rain Shelter creating Rest Area near Dragon Boat Sculpture
    鄰近龍舟雕塑設雨亭的休憩廊
(6) Dragon Boat Sculpture at Wavy Lawn
    設於波浪形狀草坪的龍舟雕塑
(7) Splashy Plaza with Playful Planter Seating Features and Granite
    Paving in Fish-Scale form enhances the Marine Image of the Park
    呈波浪形配置的座椅、水滾形的真樹和魚鱗形地磚增設的濺廣
(8) Entrance at Fuk Man Road
    福民路公園入口
(9) Wooden Trellis in the form of Ship's Hull with Seating underneath
    船形木葺及座椅
(10) Elderlies' Exercise Equipment
     長者健體設施
(11) Existing Trees along the Site to be Retained
     保地的樹木將被保留
(12) Newly Constructed Bus Stop
     新建設之巴士站

Po Tung Road

Public Transport Interchange

Sai Kung Police Station

Fuk Man Road

Sai Kung Town Hall

Urban Park - "Sai Kung on Marina"
市鎮公園 – "西貢遊艇停泊公園"

Figure 11.2 Outline scope of works for the NEC pilot case study project (Chan et al., 2014) (with permission from Sweet & Maxwell).

Table 11.3 Final out-turn performance of the project (Chan et al., 2014) (with permission from Sweet & Maxwell)

KPI	Target outcome	Final out-turn performance
1 User satisfaction	7 out of 10[1]	• 8 out of 10 by the Leisure and Cultural Services Department and by the Drainage Services Department • 8 letters of commendation received
2 Time performance	24 weeks earlier completion (i.e. completed ahead of schedule by 14.5% or 6 months)	• Completed in 141 weeks against an extended contract period of 165 weeks (i.e. completed ahead of schedule by 24 weeks or 6 months)
3 Cost performance	5% cost saving for gain share between the employer and the contractor	• Agreed final target cost HK$76.7 million less actual project cost HK$72.9 million at completion = gain share of HK$3.8 million (i.e. actual project cost reduced by HK$3.8 million, equivalent to about 4.95% cost saving from the agreed final target cost)
4 Quality performance	No major rework required	• Minor defects found on non-critical items. Handover to Leisure and Cultural Services Department within 10 days of completion of works
5 Safety and environmental performance	Zero accidents	• No recorded incidents of safety or environmental infringements, and 4 valid public complaints • 3 industry safety awards

Note:

1 Score range 1–10; 1 = totally unsatisfactory; 10 = totally satisfactory.

In the first partnering workshop, the project team, involving the representatives from the employer's (government) departments, the consultant team and the main contractor, jointly listed and agreed a set of five key performance indicators in order to pursue outstanding performance for this project, the first one adopting NEC3 in Hong Kong.

As shown in Table 11.3, generally speaking, the outcome of this project was satisfactory. Regarding user satisfaction, the two employer departments (i.e. the Leisure and Cultural Services Department and the Drainage Services Department) rated it as eight out of ten marks. This result obviously illustrated that the employer departments considered that the project was outstanding generally. The evaluation may justify this approach.

Motives for using TCC

Cost reduction

The builders opined that they would commit more human resources during NEC tendering, varying from 10% to 30% (in terms of person-hours), than in the conventional contracts adopted in civil engineering works in Hong Kong. In terms of the cost of resources transferred, the builders also expressed that they would commit more cost for tendering, varying from 5% to 30%, for this project when compared with those procured by the general conditions of contract commonly used by the client.

Hughes et al. (2012) stated that cost performance is considered the most significant success factor in projects under TCC. This open nullah improvement project saved HK$3.8 million, around 4.95% of the agreed final target cost (Chan et al., 2010a). Therefore, the cost performance was basically satisfactory since cost savings were achieved.

Improving subletting arrangements

One of the obvious reasons for this was that the NEC project utilised activity schedules instead of bills of quantities, and this option certainly contributed to aggravating the difficulties encountered with subcontracting arrangements for the main contractors. Another probable reason was that using NEC was still new for the builders, so they likely needed more time and effort to fill in the contract data and set a bidding strategy. In future, it might be helpful for the client to invite qualified tenderers to a builders' forum or a tender briefing to clarify and discuss any special matters needing careful handling, such as the client's requirements, type of contract adopted, appropriate tender submissions, suitable pricing documents, when tendering for projects adopting the NEC3 ECC.

Perceived benefits of adopting the NEC3 ECC

The contractors' perceptions concerning the benefits of using NEC3 when compared with traditional contracts are shown in Table 11.4. The respondents realised the benefits of NEC3 over traditional contracts. In this section, only those benefits mentioned by two or more respondents are discussed (shown in bold in Table 11.4). For further details of this study, please refer to Chan et al. (2014).

Simple and clear language

Two of the major objectives of NEC3 contracts are simplicity and clarity of contractual language. The lack of legal jargon in the contract is obvious

Table 11.4 Perceived benefits of using NEC3 ECC over traditional civil engineering works contracts (Chan et al., 2014) (with permission from Sweet & Maxwell)

Perceived benefits of NEC3 ECC	Contractor 1	Contractor 2	Contractor 3	Contractor 4	Contractor 5	Contractor 6	Total no. of hits
1 Simple and clear language	✓	✓					2
2 Providing financial incentives for contractor to save project costs and reduce disputes	✓			✓			2
3 Allocating risks more fairly and promoting co-operation due to open-book accounting regime	✓						1
4 A non-adversarial partnering arrangement				✓		✓	2
5 Focusing on management and actions people take as much as the obligations and liabilities of the parties				✓			1
6 Inherent recognition of the dynamic nature of civil engineering works				✓			1
7 All parties achieve common goals to speed up the works					✓		1
8 Encouraging joint problem resolution				✓	✓	✓	3
9 Improved predictability of programme, price, quality and safety					✓		1
10 Better security of supply chain					✓		1
Total no. of perceived benefits identified from each respondent	3	1	3	2	3	2	

compared with the traditional civil engineering works contract used by the HKSAR Government. The findings regarding the benefits of NEC3 in both New Zealand and United Kingdom are further reinforced by this perception (e.g. Lord et al., 2010; Wright and Fergusson, 2009). Lord et al. (2010) further pointed out that the clarity of the roles of the different contracting stakeholders can be enhanced by the use of simple language and guidance notes.

Providing financial incentives for contractors to save project costs and reduce disputes

Financial incentives are provided by the contract for contractors to illustrate their best attempts to finish the work and reduce project costs by contributing their expertise, creative ideas and alternative solutions in both construction methods and design in a rational manner. The NEC methodology did facilitate the initial involvement of the contractor to utilise its expertise in different building technologies, material selection and programming at the design stage. Time and quality performance could thus be guaranteed. Consequently, the contractor could obtain an acceptable profit margin and was likely to share any cost savings with the employer under the established gain-share/pain-share mechanism. The contract might also lead to a reduction in disputes because of the use of simple language and because the classification of compensation events was set out clearly out in the contract clauses.

Since the classification and valuation of diverse potential changes had been agreed gradually during the course of construction, the occurrence of contentious issues and uncontrollable disputes was significantly decreased. Moreover, the preparation and agreement of the final project account could be completed earlier than with the traditional approach.

Non-adversarial partnering arrangement

With reference to the Technical Circular of the Environment, Transport and Works Bureau (2004) (now re-named the Development Bureau), the project delivery technique – partnering – could be related to various forms of contracts. Partnering was adopted in the project, and this probably helped to develop a co-operative working relationship between the client and the contractor. The NEC form of contract encourages the use of a partnering strategy among the major contract parties by implementing a more collaborative and less controversial or litigious approach. Chan et al. (2010c) and Meng (2012) had a similar view that the important elements (e.g. effective communications, mutual trust and harmonious culture) help to enhance the overall performance of projects under TCC.

Encouraging joint problem resolution

Joint problem resolution by the early warning mechanism is encouraged by the harmonious partnering working relationship under NEC3 ECC. The partnering spirit helps the stakeholders to achieve common goals, and creates a teamwork culture enabling them to resolve disputes or problems together without reverting to prolonged litigation. In addition, TCC links the individual interests of the client and the builder, so the two parties are inclined to solve problems together to acquire the most desirable outcomes for the project. The study of Zimina et al. (2012), which discussed two case studies, supported this view. It illustrated that TCC can lead to a more co-operative decision making process compared with traditional procurement strategies. The project stakeholders work in the same site office to further catalyse more prompt but informal broad-minded communications, in order to enable productive joint problem resolution in a teamwork environment.

Major difficulties in implementing NEC3 ECC

Table 11.5 outlines a list of major difficulties experienced by the contractors compared with traditional contracts (Chan et al. 2014). Only those difficulties

Table 11.5 Major difficulties in implementing NEC3 ECC when compared with traditional civil engineering works contracts (Chan et al., 2014) (with permission from Sweet & Maxwell)

Major difficulties of NEC3 ECC	Contractor 1	Contractor 2	Contractor 3	Contractor 4	Contractor 5	Contractor 6	Total no. of hits
1 Intensive use of resources due to demanding requirements of document administration and record keeping	✓	✓	✓		✓		4
2 Some grey areas about contractor's responsibilities in tender document	✓						1
3 The enforceability of the partnering clause is questionable	✓						1
4 The contractor's share is pre-fixed by the employer under the contract, and the formula stated in contract data cannot produce a commercially acceptable reward to contractor		✓					1
5 **Difficulty in subletting due to complicated approval process**		✓				✓	2
6 Incompatibility with existing public procurement rules			✓				1
7 Lack of familiarity with or misunderstanding of NEC methodology by industrial practitioners in Hong Kong			✓				1
8 Difficulty in subletting works without the provision of bill of quantities				✓			1
9 Longer period of tendering process compared to traditional contracts				✓			1
10 Higher risk for bidding target cost without the provision of bill of quantities				✓			1
11 Potential failure to adopt cultural change necessary to ensure proper use of NEC						✓	1
Total number of major difficulties identified from each respondent	3	3	3	3	1	2	

reported by two or more respondents are examined in this section (shown in bold in Table 11.5).

Intensive use of resources due to demanding requirements of document administration and record keeping

One of the major objectives of the NEC is to promote desirable project management by consolidating contractual procedures with specific timeframes. On the other hand, the workloads of site staff and managerial personnel increase in terms of record keeping and documentation. For instance, keeping receipts or invoices for each item is necessary for auditing purposes under the open-book accounting arrangements of this project. As a result, every stake-holder (client, contractor and consultant) needs to increase its staff resources.

Difficulty in subletting due to complicated approval process

The subcontracting process was complicated by the complex approval procedures under the NEC. This problem was mainly ascribed to the procurement process stated in the government's Store and Procurement Regulations. Enormous pressure was placed on the builders in terms of subletting issues, which affected the progress of the programme.

Critical success factors

The partnering workshop facilitator carried out semi-structured face-to-face interviews with key project stakeholders from the employer and contractor to investigate the success factors behind this case. The representatives submitted their tender proposals and gathered essential information and data from the close-out partnering workshop.

After the interviews, the success factors were summarised by Chan et al. (2014) as follows:

- There was strong leadership by senior management of the client, supervisor and contractor in linking individual objectives to the overall project goals and supporting collaborative working relationships within the project team.
- A mutually agreed set of target outcomes was established against which the project team could measure and drive performance.
- There was appropriate selection of a project management team with co-operative working attitudes, and this team was aligned and developed by means of regular and effective facilitated workshops and meetings.
- The client in the role of project manager was more closely involved and had greater responsibility for the timely resolution of any disputes or conflicts compared with traditional contracts.
- The project manager made fair decisions in the best interests of project performance.
- There was facilitation of and support for the frontline working team in converting from an adversarial relationship to a co-operative one.
- The government stakeholders and local community were engaged during the development of designs and reduction of environmental impacts affecting the public.
- Open and early formal notification of defects left little room for design or site errors to be tackled informally.

It is evident that most of these items are relationship-based, such as closer involvement of the employer and stronger support from senior management of the key players. NEC3 integrates obligations with project management processes and encourages co-operation between client and contractor. It is obvious that this project sets a benchmark for future projects in Hong Kong.

GMP: two case studies from South Australia

Background

GMP projects in South Australia

Perkin and Ma (2010) summarised the current and completed projects in the construction sector in South Australia that have adopted the GMP strategy, as shown in Table 11.6. Those projects were identified through an extensive search of active industry organisational websites. The construction periods varied from 40 to 144 weeks, and the contract values ranged from AU$4.4 million to AU$84 million. The procurement methods under the GMP contract include negotiated, design-and-build and fixed lump-sum, which are popular delivery systems. It is worth noting that though the GMP approach was introduced in the South Australian construction industry in the early 1990s, most GMP projects have been completed during 2004–2008. This shows a strong trend towards the adoption of the GMP procurement method.

Two anonymous projects in South Australia that adopted the GMP approach showed desirable outcomes.

A private commercial building in South Australia (Project A)

Project A was a unique and landmark commercial project. The contracting parties had a reasonable level of experience with the GMP procurement

Table 11.6 List of GMP projects in South Australia (Perkin and Ma, 2010)

Project title	Delivery system	Project nature
Central City Bus Station	Design & Construct	Commercial, residential
Munno Para Shopping Centre	Negotiated GMP	Commercial
SA Water Building (VS1)	Design & Construct	Commercial
SA Water Fit-out	Design & Construct	Commercial
Liberty Tower Apartments	Design & Construct	Residential
151 Pirie Street	Design & Construct	Commercial
Paradise Wirrina Cove	GMP	Commercial
Burnside Village	GMP	Commercial
Hirotec Automotive Closure Plant	GMP	Industrial
Rendezvous Allegra Hotel	GMP	Commercial
Wine Export Facility	Negotiated GMP	Industrial
Wayville Showgrounds	Design & Construct	Recreational, special use
Harvey Norman	Design & Construct	Industrial, commercial
Gepps Cross – Home HQ	Design & Construct	Commercial

approach. A GMP contract was suggested by one of the employer's partnering consultants at the detailed design stage. The method of tendering was selective without partnering or a pain-share/gain-share agreement involved in the procurement stage. If funding was obtained and approved, because of the limited timeframe of the project, an alternative design and construction method would be implemented. The tender documents would be open to the market for pricing when they were at 70% completion. When the contractor was chosen, the design consultants were novated to the contractor and the design would be completed.

A private industrial building in South Australia (Project B)

Project B was procured as GMP of an industrial nature without partnering. The method of tendering was selective when the documents were about half-completed, which was recommended by the employer's team at the feasibility stage of the project.

The documents included a thorough principle project requirements document, architectural drawings, structural drawings and services drawings (all about half-completed) and a contract works specification (half-complete). An open-book approach with a gain-share/pain-share mechanism was specified in the contract documents, stipulating that once every subcontract was let, it would be reviewed and monitored by the employer, with the savings or extra costs being shared between the employer and contractor.

Generally, both of the projects were considered to be successful, as Project A was completed on schedule and within the employer's budget and the outcome of Project B was acceptable to the employer. Perkin and Ma (2010) commented that the partnering approach was more likely to lead to greater success for all parties.

Table 11.7 Underlying motives for Project A and Project B (Perkin and Ma, 2010)

	Project A	*Project B*
Similarities	• Better working relationship • To capitalise on the builder's expertise in design • To enhance risk management and control • More time saving due to the overlap of design and construction	• Better working relationship • To capitalise on the builder's expertise in design • To enhance risk management and control • More time saving due to the overlap of design and construction
Differences	• Past desirable experience with GMP projects • To pre-agree a ceiling price and time at contract award/commencement	• To improve the quality of the built facilities • The need for an open-book accounting arrangement • To develop an incentive to achieve cost savings

Table 11.8 The rationale behind using GMP/TCC (Perkin and Ma, 2010)

Rationale	Case Study A	Case Study B
Tendering method	Selective	Selective
Contracting approach	Novated design-and-build at 70%	Design-and-build at 50% (first cut GMP)
Gain-share arrangement	No arrangement	50:50 arrangement
Pain-share arrangement	No arrangement	70:30 arrangement (client:contractor)

Motives and rationale for using GMP/TCC

Motives

The underlying motives for Project A and Project B are shown in Table 11.7.

Rationale

The rationale for both projects is shown in Table 11.8.

Lessons learned

Project A

Perkin and Ma (2010) suggested that holding a workshop involving all stakeholders could help to manage some major issues of the project. They also recommended that for selective tendering, all tendering parties should take part in a session to comprehend the employer's expectations and requirements, to enhance understanding of the project. In addition, GMP should be determined subsequent to the engagement of the builder, and selective tendering should be applied through the managing contract, which required preliminaries, overheads and margin only in the tenders. Last but not least, the consultants should participate in the pain-share/gain-share arrangement, as it may encourage them to achieve a common goal with the employer and builder.

Project B

The respondents in the study of Perkin and Ma (2010) recommended that the risks should be transferred to the party best able to handle them by establishing a comprehensive risk register and risk management system (risk assessment tools). A risk management workshop could be held regularly and involve all stakeholders (open forum). It was suggested that a partnering approach be applied. However, in Project B, because of the nature of the project and insufficient ability to adopt partnering, this strategy could not be employed.

Overall

In general GMP projects, employers tend to transfer more risks to the builders, but they do not want to pay for the risks, so the builders increase their tender prices to cover the risks

Regarding the documents, it was recommended that they should be completed by the employers. If the contractors are responsible for those documents, they may increase the tender price and provide savings at a later stage.

However, the pain-share arrangement was regarded to be inappropriate for design-and-build GMP projects, since it motivates builders to be claim-conscious.

In addition, the definition of a successful project and the critical success factors should be established so that they can be monitored throughout the project.

GMP: Chater House, Hong Kong

Background

Chater House is a private office development containing twenty-nine floors with total gross floor area of 74,000m² in the central business district of Hong Kong, and it accommodates high-end retail outlets on its lower part to international Grade A standard. The construction phase consisted of the demolition of the existing building, and the construction of foundations and the office tower section. The actual project cost was about HK$1.5 billion. The final project duration was 635 calendar days. The project utilised the GMP procurement method with a gain-share mechanism and negotiated tendering.

This was the second project in Hong Kong to use the GMP strategy after the success of the 1063 King's Road project. Although it was constructed during a serious economic downturn in Asia, the developer decided to continue the development of Chater House. Since one of the objectives of this project was to redevelop the area into a prestigious commercial office with luxury retail outlets, the developer chose to implement an alternative integrated procurement arrangement to develop an innovative and constructive working environment to achieve the objectives.

Moreover, set against a backdrop of declining standards in local construction and an adversarial working relationship between the employer and builder associated with indifference towards progressive thinking, the employer on the Chater House project intended to establish an environment that would encourage the development of creative ideas, a collaborative relationship between the contracting parties and new technologies, with the objective of achieving the elusive win–win result and desirable quality, reducing construction waste and improving safety standards for workers

(HK-BEAM, 2005). Due to time and efficiency initiatives, the project was finished six days early and the actual project cost was decreased by around HK$27 million (approximately £1.93 million), which equals a 15% cost saving compared to the initial budget (Uebergang et al., 2004). Based on a contractor's internal data and records, the construction waste resulting from the Chater House project was 25% lower compared with those using conventional construction strategies for office buildings of a similar scale (D.W.M. Chan et al., 2011b). The arrangements of GMP helped the project come to completion earlier than the stated schedule, with cost savings, higher quality, fewer disputes, a more collaborative working relationship and less material wastage.

Motives for using GMP

Transparency and good working relationship

J.H.L. Chan et al. (2011a) found that the conventional forms of building contract were reviewed and discounted as being poorly suited to the transparent and open working relationship fostered by the employer. A contractor emphasised that 'the project client intended to follow a procurement route that complemented the partnering strategy'. Since a hybrid contract relied on the standard negotiated form but imposed a price cap and a fixed completion date, the GMP contract had been implemented successfully for the employer's two previous private building projects (1063 King's Road, Hong Kong and One Raffles Link, Singapore). Therefore, the operational mechanism of GMP procurement was advocated, encouraging the prime contracting bodies (the employer, main contractor and consultants) to work as a team in deciding the method of construction, time and cost management, a detailed breakdown of direct works, preliminaries and conditions of contract (Chan et al., 2004a). The open-book accounting arrangement meant that the main contractor should make all its backup cost data available to other project stakeholders. Exchange of information required mutual trust among the contracting parties, especially the main contractor.

Encouraging collaboration

The employer expected that the GMP procurement approach with implementation of partnering would encourage further collaboration between the employer and the main contractor. This common objective and the open-book accounting arrangement cultivated a sense of partnership and mutual trust among the contracting parties. Moreover, regular partnering review meetings and the adjudication committee operated under the GMP procurement method set up an ideal mechanism to tackle any problems encountered and deal with confrontational matters (Chan et al., 2003). In addition, the employer followed the direction of the GMP, a 'co-operative contracting' approach, which achieved the following objectives (D.W.M. Chan et al., 2011b):

- to obtain a competitive price;
- to exercise control over the design and construction processes;
- to speed up the construction phase;
- to maximise value for money;
- to achieve a level of quality in line with the rest of client's portfolio and its expectations for the new building;
- to facilitate a transfer of risk and a sharing of reward with the main contractor.

Major benefits of adopting GMP

Chater House outperformed other similar projects under traditional contracts. D.W.M. Chan et al. (2011b) interviewed five project stakeholders, including the client, consultant and main contractor. They pointed out the following perceived benefits from applying GMP approach.

Achieving a competitive price

All interviewees agreed that the GMP strategy could achieve a competitive price, value for money and high quality of works, and provided stronger incentives for innovation and cost saving via the gain-share mechanism.

At the tender stage, all nominated and domestic subcontract works packages were tendered in a keen competition, the former being mainly managed by the employer's consultant team, and the latter by the main contractor. A prevalent thread running through the procurement process was the 'round-table' agreement by all stakeholders on which subcontractors should be invited to tender for the work and further subcontracts. A representative from the employer's side opined: 'this "open-book" tendering process ensured that the client received competitively priced tenders from approved subcontractors and specialists'. Since the subcontractors' tenders had to be approved by employer, the employer could retain control over the work package contractors to be selected for a number of works packages. By adopting an open-book accounting strategy, responsibility and quantification of the costs of risk were improved, according to the National Economic Development Office (1982). Concurrently, retaining control over the design stage prevented the contractor from adopting a lowest-cost approach, which was contrary to a value-for-money approach, to increase its share of potential cost savings.

Providing incentives for cost and time savings

In GMP contracts, the gain-share mechanism was another distinctive characteristic (Trench, 1991). If savings resulted from the difference between the final contract sum and the agreed GMP, the employer and contractor could share the gain according to a pre-agreed ratio. Not only were the individual interests of different stakeholders integrated, this strategy also helped achieve 15% project cost savings. The main contractor expressed that this approach in fact generated a strong motivation for contractors to achieve cost savings by integrating the contractor's experience, knowledge and innovations in both design and construction methods. GMP procurement strategies enabled early involvement of the builders to contribute their buildability expertise on other construction techniques and materials throughout the design stage. Therefore, performance in terms of time and quality could be guaranteed.

Encouraging innovations

Forty-seven innovations were introduced to the project development. For example, the typical floor construction utilised an integration of self-climbing hydraulic steel formwork for the central core and aluminium table formwork for the post-tensioned slabs. Both of the formworks were new in Hong Kong, and the latter one was designed, supplied, installed and operated by the main contractor. These two formwork techniques greatly enhanced the efficiency of production of the structural frame compared with the conventional formwork system. This improved the quality of the finished concrete in terms of finish and construction tolerances, resulting in less wastage. Compared with traditional methods, D.W.M. Chan et al. (2011b) found that 25% less waste was generated during the Chater House project.

Proper risk sharing

The main contractor opined that the GMP strategy led to a mutually acceptable risk sharing and reward mechanism. The contractor took into account the design development risk in terms of GMP allowance in the tender, and the employer was offered the comfort of a 'not-to-exceed' contract sum in the early project development process, which could only be modified as a result of either scope changes or adjustments to provisional quantities or provisional sums. It enabled a certain level of flexibility on cost by offering a design development fund for miscellaneous variations, concurrently alleviating the financial risks taken by the employer (Boukendour and Bah, 2001). In addition, a representative from main contractor pointed out that:

> the GMP form of contract was conducive to implementing partnering approach into the working relationships between various key project stakeholders, with the objective of achieving the 'maximum price' by adopting a more co-operative and less contentious or litigious approach to the contract.

A quantity surveying consultant also opined that:

> as the valuation of variations must be agreed progressively during the construction phase, the occurrence of disputes was greatly reduced and the preparation and agreement of the final project account was finalised earlier than an average traditional lump-sum contract.

Ability to manage serious problems

The significant benefit of the GMP contracts in this project was the capability to tackle severe problems (e.g. late design changes) by means of the open-book accounting arrangement and the co-operative working relationship via partnering. In this project, due to the recession of the office rental market in 2000, the architects were requested to review the design, which had already approved by the Buildings Department. Chater House was re-designed and adopted a more regular shape in order to reduce costs while providing a footprint that tenants preferred (Tam, 2002). 'This usually would result in significant problems and delay to the project,' a client representative commented, 'however, under the GMP contractual arrangement, these issues were resolved in a timely and efficient manner on account of the effective project management, partnering spirit and proper decision making methodologies adopted throughout'.

Major difficulties in implementing GMP

Definition of scope of work leading to disputes

As stressed by most interviewees, the fundamental problem in adopting the GMP procurement method for the Chater House project was definition of the scope of work, resulting in intractable disputes. Gander and Hemsley (1997) and Fan and Greenwood (2004) shared a similar view with the interviewees. Unexpected changes were common during construction, and the price required adjustment of the agreed GMP value. In addition, the degree of design development variations was difficult to identify. If these issues are not managed appropriately, it may lead to intractable disputes and damage the established mutual trust among the stakeholders, which was also the concern of Chevin (1996) and Chan et al. (2010b). Therefore, the GMP approach may not be a suitable strategy for contracts where numerous changes are anticipated or it is difficult to define the scope of work at the early stages (Trench, 1991).

Difficulty in evaluating the variations

When unexpected changes appeared, it was natural that the employer would pull in an opposite directions to the builder to achieve its own objectives.

Although GMP variations were defined in the methodology document, the builder tended to regard the variations as a scope change to increase the final savings, while the employer prefer to cover those costs with the design development allowance to maximise the value of the works even if potential cost savings were sacrificed. Tang and Lam (2003) also viewed that it was difficult to assess the revised contract price when the contractor proposed an alternative design, taking more time to re-evaluate the cost implications. The successful team building and mutual trust developed drove the performance of the project. Moreover, the tender briefing and the dispute resolution mechanism needed to be comprehensive, transparent and equitable to reduce this risk.

Higher tender price

One main limitation inherent to the GMP procurement method was the financial risks for both employer and contractor, since there were some uncertainties relating to the scope of work. The representative from the contractor suggested that:

> compared with the conventional procurement method, not only that the contractor had to bear risks in both the design and construction processes, his risks were further inflated for a GMP project due to the 'guaranteed price' but with the absence of pain-share mechanism.

The builder tended to inflate the tender sum to cover additional risks, as additions or changes in the scope of work could only be claimed if they were classified as GMP variations. Therefore, assessment and negotiation were essential to achieve a sensible and mutually agreed GMP and provisional sum to achieve successful project delivery.

Unfamiliarity with GMP contracts

Since the GMP method was still in its infancy in Hong Kong, with few cases of application, stakeholders tended to be unfamiliar with it and disapprove of the GMP procurement arrangement, as expressed by all interviewees. Although structured procedures were stated in GMP scheme, the consultant team was sceptical of this procurement arrangement when drafting the contract with the employer. A client representative opined that if variations were frequent without limiting the appropriate design development allowance, or if there were serious disputes between builder and employer, the project delivery of GMP contract might not be successful. If subcontractors did not thoroughly comprehend the philosophy of GMP, but nevertheless accepted this procurement arrangement, the implementation of GMP contract might be adversely affected, leading to failure of the entire project.

Critical success factors for GMP

Willingness to achieve co-operation

The interviewees believed that the genuine willingness to collaborate helped achieve the project's success. This result was also supported by Tay et al. (2000), who found that a close working relationship or partnering spirit among all project stakeholders was crucial for facilitating a project. A structured partnering mechanism, as illustrated in Figure 11.3, was adopted as a complementary strategy to the GMP contract, including regular partnering workshops throughout the construction stage, nomination of partnering champions, a conflict resolution mechanism, a partnering performance monitoring system, and a feedback system at completion. This partnering arrangement allowed the project stakeholders to express their views and concerns freely. It served to enhance mutual trust, facilitate effective communication and improve working relationships among the contracting parties in order to achieve common objectives (Chan et al., 2004). The gain-share approach and open-book accounting strategies under GMP procurement could also enhance the development mutual trust among the project stakeholders.

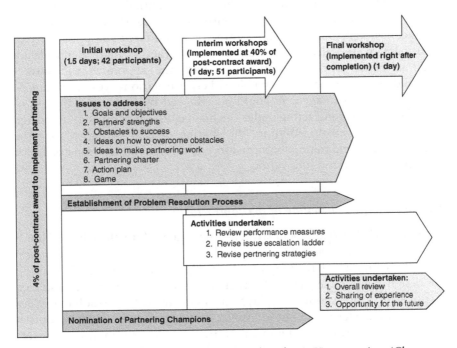

Figure 11.3 Partnering approach and process for Chater House project (Chan et al., 2004) (with permission from American Society of Civil Engineers).

Establishment of an effective adjudication arrangement

Most of the interviewees opined that an effective adjudication process and adjudication committee were crucial to the GMP strategy. The adjudication committee would report on the status of a variation submission and determine the categorisation of different variations requested by the builder, i.e. whether the variation was classified as a design development (cost attributed to the agreed design development budget and thus having no influence on the agreed value of the GMP) or a scope change (a substantial change to the overall function, floor plan area, quality or quantity of an area which could change the agreed GMP value). Moreover, it is essential that mutual agreement on the valuation of variations be achieved as promptly as possible to prevent any impact on the overall progress of the project. The quantity surveying consultant should be impartial in order to achieve success.

Appropriate arrangements for subletting process

The subcontract procurement process was essential to achieving the employer's expectations in terms of quality performance. 'The project was procured based on an "open-book" accounting arrangement with joint tendering and selection of subcontractors,' a representative from the contractor stated, and 'the main contractor was offered an opportunity to participate in the selection process of subcontractors that would ideally result in better working relationships with the appointed subcontractors, and resolving interface omissions between various works packages.' Efficient project management could be achieved by preventing the use of multi-layered subcontracting under the main contract document unless the employer approved. In addition, the sharing of cost savings among the main contractor and trade subcontractors could drive construction excellence through efficiency and innovation.

Allowing early participation of the contractor

Under the GMP procurement strategy implemented for the Chater House project, the main contractor team could be involved at the early design stage. It is important to integrate the competence of the builder during the design stage under TCC. Mutual trust can be smoothly transferred to site level. Correspondence plays an important role in resolving issues promptly at site level, but not following contractual procedures. Adopting technical innovations by the main contractor can help minimise project time and cost. The concepts of GMP at the outset of project development can thus allow early involvement by the builder in both construction methods and design.

TCC: London Heathrow Airport Terminal 5

Background

London Heathrow Airport Terminal 5 was one of the most complicated and largest infrastructure projects in Europe, involving sixteen main interconnecting projects and fourteen sub-projects (EC Harris, 2009; Potts and Ankrah, 2014). It had a value of £4.3 billion and was delivered on time and on budget.

TCC helped this high-profile project to achieve success (Williams et al., 2013). After completion, it added 50% to the capacity of Heathrow and offered a spectacular gateway to London (Potts and Ankrah, 2014).

TCC: Hydro power station at Lake Manapouri, New Zealand

Background

New Zealand

Wright and Fergusson (2009) found that the electricity industry is mature in New Zealand, but it has been divided into generation, transmission, distribution and retailing to various industry sectors. This was driven by government initiatives to develop a competitive market with higher efficiency (Ministry of Business Innovation and Employment, 2012). In the past six years, the broadly capable electrical engineering service providers have reduced to three. Market competitiveness decreased and the providers' bargaining power increased, since the industry sectors all compete for constrained supply of services.

Due to the increasing demand and reduced competition of supply, contractors are less motivated to work co-operatively with clients and are encouraged to adopt an opportunistic strategy of simple revenue maximisation at the clients' cost (Wright and Fergusson, 2009).

Historically, Meridian adopted conventional forms of engineering and construction contracts for engineering works, generally the FIDIC (Fédération Internationale Des Ingénieurs-Conseils) form. From 2002, Meridian has tried to convert traditional adversarial-styled contracts into a more collaborative working style which was characterised by Turner and Simister (2001) as 'relational contracts', suitable for well-defined contracts, focusing on efficiency, and operating with trust and mutual respect.

In fact, the initial attempts were unsuccessful since the builders were reluctant, sceptical of the perceived benefits, and had no commitment to or interest in changes.

Hydro power station at Lake Manapouri

Despite unsuccessful experiences, Meridian has awarded two contracts to refurbish and upgrade the largest hydro power station in New Zealand, which is located at Lake Manapouri. This project used two contracts. The first was the 1999 edition of the FIDIC contract for the electrical works. The second one was the second edition of the NEC ECC for the mechanical works.

The electrical and mechanical contractors were required to upgrade seven generators, but they had to ensure that only one unit was out of service at any time. Before the mechanical contractor started work, the electrical contractor had worked on three generators. Both contractors subsequently worked at the same time on the next four units (Wright and Fergusson, 2009).

Perceived benefits of NEC ECC

Wright and Fergusson (2009) set twenty-one open-ended questions regarding clarity, cost, time and relationships. They found that ECC provided:

- a structured project management framework with terms and conditions which focus the client's and contractor's attentions on the performance of the project rather than the boundaries stated in the contract;
- flexibility in its terms and conditions due to its range of standard bolt-on options, and flexible payment options determining how the financial risks will be allocated;
- well-defined contract procedures, contributing to the efficiency of both the contract and project management;
- clear and simple language, avoiding legalese and ambiguity, and enhancing understanding; when uncommon terms are included, the instances are minor and cause no difficulty once users become familiar with them;
- well-defined contract roles and clear allocation of responsibilities and risk among the project stakeholders, with flexibility to transfer the risk to the party best able to handle it;
- a proactive and forward-looking approach with requirements to continually plan in order to achieve the project outcome;
- higher costs than conventional contracts because of the increased daily management effort; nevertheless, these costs could arguably be offset by reducing the potential costs of resolving problems and settling disputes, and by the requirement for stakeholders to work collaboratively to resolve problems before they become serious; the early warning arrangement can provide efficient management of turn-out costs;
- motivation for collaboration among project parties to work together in a transparent and honest relationship, empowering the client and builder to communicate directly without passing through a third-party engineer-to-contract, preventing game-playing and disputes, and encouraging an environment of goodwill among the stakeholders.

Table 11.9 Summary of case studies

Projects	London 2012 Olympic Park Development	MTRC Tsim Sha Tsui Metro Station Modification Works	Open Nullah Improvement Project in Sai Kung	Case A	Case B	Chater House	London Heathrow Airport Terminal 5	Hydro Power Station at Lake Manapouri	Frequency
Locations	London	Hong Kong	Hong Kong	South Australia	Hong Kong		London	New Zealand	
Contract type	TCC	TCC	TCC	GMP	GMP		TCC	TCC	
1. Motives behind using TCC/GMP									
Develop better working relationship	✓	✓		✓	✓	✓			5
Generate financial incentives		✓	✓		✓	✓			4
Achieve time savings		✓		✓	✓	✓			4
Improve risk management	✓			✓	✓	✓			4
Provide high degree of transparency	✓				✓	✓			3
Utilise contractor's expertise in design				✓	✓				2
Past successful experience			✓	✓					2
Common adoption	✓								1
Improve subletting arrangements			✓						1
Set up a ceiling price				✓					1
2. Perceived benefits of TCC/GMP/NEC									
Improve working relationship	✓		✓					✓	3
Provide equitable risk apportionment			✓			✓		✓	3
Establish mutual objectives	✓	✓	✓						3
Involve client in problem solving and subletting			✓			✓		✓	3
Provide financial incentives		✓	✓						2
Standardisation of documentation	✓							✓	2
Simple language			✓					✓	2
Achieve better value for money			✓			✓			2

(continued)

Table 11.9 (continued)

Projects	London 2012 Olympic Park Development	MTRC Tsim Sha Tsui Metro Station Modification Works	Open Nullah Improvement Project in Sai Kung	Case A	Case B	Chater House	London Heathrow Airport Terminal 5	Hydro Power Station at Lake Manapouri	Frequency
Locations	London	Hong Kong	Hong Kong	South Australia	Hong Kong	Hong Kong	London	New Zealand	
Save time			✓						1
Utilise contractor's expertise in design						✓			1
Reduce disputes								✓	1
More control by clients		✓							1
Early warning mechanism	✓								1
3. Major difficulties in implementing TCC/ GMP/NEC									
Require many sources in document administration and record keeping	✓		✓						2
Lack of confidence/ experience		✓				✓			2
Disputes from determining the types of variations		✓				✓			2
Disputes from unclear definition of scope of work						✓			1
Absence of standard forms of contract		✓							1
Difficulty in subletting			✓						1
Longer period of tender process			✓						1
High risk if without bill of quantities			✓						1
Potential failure due to cultural changes			✓						1
Higher tender price						✓			1
4. Critical success factors									
Ensure every party agree the programme at the outset	✓			✓	✓	✓			4

Maintain good relationship	✓	✓	✓			✓	4
Prepare more resources	✓			✓	✓		3
Maintain good communication and commitment	✓		✓		✓		3
Right selection of project team		✓	✓				2
Ensure transparency		✓	✓				2
Acquire sufficient project information	✓						1
Ensure tenderers provide enough supporting information	✓						1
Provide sufficient incentives		✓					1
Get more involved in the resolution of disputes			✓				1
Equitable decisions by project managers			✓				1
Impartial risk allocation				✓			1
Establish an effective dispute resolution arrangement						✓	1
Establish appropriate arrangements on subletting process						✓	1
Allow early participation of the contractor						✓	1
More training							1

Lessons learned

In order to maximise the outcomes of ECC, Wright and Fergusson (2009) proposed the following recommendations in order of priority:

- more training for project teams, project management and contractors;
- greater awareness of the benefits of the ECC;
- encouraging wider future adoption of the ECC in the engineering and construction industry;
- actively promoting the collaborative working strategy of ECC and fostering co-operation by co-locating the project teams in site offices; this can enhance communication and help to develop a close working relationship among stakeholders, with limited extra cost involved.

Summary of case studies

Table 11.9 summarises all the case studies in this chapter, to compare performance under TCC and GMP contracts in terms of motives, perceived benefits, major difficulties and critical success factors.

References

Abednego, M.P. and Ogunlana, S.O. (2006) Good project governance for proper risk allocation in public–private partnerships in Indonesia. *International Journal of Project Management*, 24(7), pp.622–634.

Abraham, I.L. Manning, C.A., Snustad, D.G., Brashear, H.R., Newman, M.C. and Wofford, A.B. (1994) Cognitive screening of nursing home residents: factor structures of the Mini-Mental State Examination. *Journal of the American Geriatrics Society*, 42(7), pp.750–756.

Abrahamson, M. (1984) Risk management. *The International Construction Law Review*, 1(3), pp.241–264.

Adams, F.K. (2008) Risk perception and Bayesian analysis of international construction contract risks: the case of payment delays in a developing economy. *International Journal of Project Management*, 26(1), pp.138–148.

Ahmed, S.M., Ahmed, R. and De Saram, D.D. (1998) Risk management in management contracts. *Asia Pacific Building and Construction Management Journal*, 4(1), pp.23–31.

Ahmed, S.M., Ahmad, R. and De Saram, D.D. (1999) Risk management trends in the Hong Kong construction industry: a comparison of contractors and owners perceptions. *Engineering, Construction and Architectural Management*, 6(3), pp.225–234.

Akintoye, A.S. and Macleod, M.J. (1997) Risk analysis and management in construction. *International Journal of Project Management*, 15(1), pp.31–38.

Al-Subhi Al-Harbi and Kamal, M. (1998) Sharing fractions in cost-plus-incentive-fee contracts. *International Journal of Project Management*, 16(2), pp.73–80.

American Institute of Architects (2001) *The Architect's Handbook of Professional Practice*, 13th edition. New York: John Wiley.

Andi, A. (2006) The importance and allocation of risks in Indonesian construction projects. *Construction Management and Economics*, 24(1), pp.69–80.

Anvuur, A.M. and Kumaraswamy, M.M. (2010) Promises, pitfalls and shortfalls of the guaranteed maximum price approach: a comparative case study. In: Egbu, C. (ed.) *Proceedings 26th Annual ARCOM Conference*, 6–8 September 2010, Leeds, United Kingdom, pp.1,079–1,088.

Ashley, D.B., Dunlop, J.R. and Parker, M.M. (1989) *Impact of Risk Allocation and Equity in Construction Contracts*. In: Uff, J. and Odams, A.M. (eds) *Risk Management and Procurement in Construction*. London: Centre of Construction Law and Management, King's College London.

Avery, D. (2006) How collaborative commercial strategies give certainty to the delivery of major railway infrastructure projects. In: *Proceedings of the PMICOS 2006 Annual Conference*, 23–26 April 2006, Orlando, FL.

Badenfelt, U. (2008) The selection of sharing ratios in target cost contracts. *Engineering, Construction and Architectural Management*, 15(1), pp.54–65.

Baloi, D. and Price, A.D.F. (2003) Modelling global risk factors affecting construction cost performance. *International Journal of Project Management*, 21(4), pp.261–269.

Bayliss, R., Cheung, S.O., Suen, H. and Wong, S.P. (2004) Effective partnering tools in construction: a case study on MTRC TKE contract in Hong Kong. *International Journal of Project Management*, 22(3), pp.253–263.

Bench, R., Webster, M. and Campbell, K.M. (2005) An evaluation of partnership development in the construction industry. *International Journal of Project Management*, 23(8), pp.611–621.

Bernhard, R.H. (1988) On the choice of the sharing fraction for incentive contracting. *Engineering Economist*, 33(3), pp.181–190.

Black, C., Akintoye, A. and Fitzgerald, E. (2000) An analysis of success factors and benefits of partnering in construction. *International Journal of Project Management*, 18(6), pp.423–434.

Bogus, S.M., Shane, J.S. and Molenaar, K.R. (2010) Contract payment provisions and project performance: an analysis of municipal water and wastewater facilities. *Public Works Management and Policy*, 15(1), pp.20–31.

Boukendour, S. and Bah, R. (2001) The guaranteed maximum price contract as call option. *Construction Management and Economics*, 19(6), pp.563–567.

Bower, D., Ashby, G., Gerald, K. and Smyk, M. (2002) Incentive mechanism for project success. *Journal of Management in Engineering*, 18(1), pp.37–43.

Bresnen, M. and Marshall, N. (2000) Building partnerships: case studies of client–contractor collaboration in the UK construction industry. *Construction Management and Economics*, 18(7), pp.819–832.

Broome, J. and Perry, J. (2002) How practitioners set share fractions in target cost contracts. *International Journal of Project Management*, 20(1), pp.59–66.

Bryant, M. (2012) London 2012 Olympic Park delivered on time and within budget using NEC3 contracts. *NEC Newsletter*, 58.

Cantirino, J. and Fodor, S. (1999) Construction delivery systems in the United States. *Journal of Corporate Real Estate*, 1(2), pp.169–177.

Carty, G.J. (1995) Construction. *Journal of Construction Engineering and Management*, 121(3), pp.319–328.

Chan, A.P.C. and Yung, E.H.K. (2003) *Procurement Selection Model for Hong Kong*, 2nd edition. Research monograph, Department of Building and Real Estate, The Hong Kong Polytechnic University.

Chan, A.P.C., Chan, D.W.M. and Ho, K.S.K. (2002) *An Analysis of Project Partnering in Hong Kong*, Research monograph, Department of Building and Real Estate, The Hong Kong Polytechnic University.

Chan, A.P.C., Chan, D.W.M. and Ho, K.S.K. (2003) An empirical study of the benefits of construction partnering in Hong Kong. *Construction Management and Economics*, 21(5), pp.523–533.

Chan, A.P.C., Chan, D.W.M., Chiang, Y.H., Tang, B.S., Chan, E.H.W. and Ho, K.S.K. (2004) Exploring critical success factors for partnering in construction projects. *Journal of Construction Engineering and Management*, 130(2), pp.188–198.

Chan, A.P.C., Chan, D.W.M., Fan, L.C.N., Lam, P.T.I. and Yeung, J.F.Y. (2008) Achieving partnering success through an incentive agreement: lessons learned from an underground railway extension project in Hong Kong. *Journal of Management in Engineering*, 24(7), pp.128–137.

Chan, D.W.M. and Chan, J.H.L. (2012) Developing a Performance Measurement Index (PMI) for target cost contracts in construction: a Delphi study. *Construction Law Journal*, 28(8), pp.590–613.

Chan, D.W.M. and Kumaraswamy, M.M. (2002) Compressing construction durations: lessons learned from Hong Kong building projects. *International Journal of Project Management*, 20(1), pp.23–35.

Chan, D.W.M., Chan, A.P.C., Lam, P.T.I., Lam, E.W.M. and Wong, J.M.W. (2006) Exploring the application of target cost contracts in the Hong Kong construction industry. In: *Proceedings of the 31st AUBEA Conference*, 12–14 July 2006, Sydney, Australia (CD-ROM).

Chan, D.W.M., Chan, A.P.C., Lam, P.T.I., Lam, E.W.M. and Wong, J.M.W. (2007a) *An Investigation of Guaranteed Maximum Price (GMP) and Target Cost Contracting (TCC) Procurement Strategies in Hong Kong Construction Industry*. Research monograph, Department of Building and Real Estate, The Hong Kong Polytechnic University.

Chan, D.W.M., Chan, A.P.C., Lam, P.T.I., Lam, E.W.M. and Wong, J.M.W. (2007b) Evaluating guaranteed maximum price and target cost contracting strategies in Hong Kong construction industry. *Journal of Financial Management of Property and Construction*, 12(3), pp.139–149.

Chan, D.W.M., Chan, A.P.C., Lam, P.T.I., Chan, J.H.L., Hughes, W. and Ma, T. (2008) A research framework for exploring risk allocation mechanisms for target cost contracts in construction. In: *Proceedings of the CRIOCM 2008 International Research Symposium on Advancement of Construction Management and Real Estate*, 31 October–3 November 2008, Beijing, China, pp.289–296.

Chan, D.W.M., Chan, A.P.C., Lam, P.T.I. and Chan, J.H.L. (2010a) Exploring the key risks and risk mitigation measures for guaranteed maximum price and target cost contracts in construction. *Construction Law Journal*, 26(5), pp.364–378.

Chan, D.W.M., Chan, A.P.C., Lam, P.T.I. and Wong, J.M.W. (2010b) Empirical study of the risks and difficulties in implementing guaranteed maximum price and target cost contracts in construction. *Journal of Construction Engineering and Management*, 136(5), pp.495–507.

Chan, D.W.M., Chan, A.P.C., Lam, P.T.I. and Wong, J.M.W. (2010c) Identifying the critical success factors for target cost contracts in the construction industry. *Journal of Facilities Management*, 8(3), pp.179–201.

Chan, D.W.M., Lam, P.T.I., Chan, A.P.C. and Wong, J.M.W. (2010d) Achieving better performance through target cost contracts: the tale of an underground railway station modification project. *Facilities*, 28(5/6), pp.261–277.

Chan, D.W.M., Chan, A.P.C., Lam, P.T.I. and Wong, J.M.W. (2011a) An empirical survey of the motives and benefits of adopting guaranteed maximum price and target cost contracts in construction. *International Journal of Project Management*, 29(5), pp.577–590.

Chan, D.W.M., Lam, P.T.I., Chan, A.P.C. and Wong, J.M.W. (2011b) Guaranteed Maximum Price (GMP) contracts in practice: a case study of a private office development project in Hong Kong. *Engineering, Construction and Architectural Management*, 18(2), pp.188–205.

Chan, J.H.L. (2011) *Developing a Fuzzy Risk Assessment Model for Target Cost and Guaranteed Maximum Price Contracts in the Construction Industry of Hong Kong*. PhD thesis, Department of Building and Real Estate, The Hong Kong Polytechnic University.

Chan, J.H.L., Chan, D.W.M. and Lord, W.E. (2011a) Key risk factors and risk mitigation measures for target cost contracts in construction: a comparison between the West and the East. *Construction Law Journal*, 27(6), pp.441–458.

Chan, J.H.L., Chan, D.W.M., Lam, P.T.I. and Chan, A.P.C. (2011b) Preferred risk allocation in target cost contracts in construction. *Facilities*, 29(13/14), pp.542–562.

Chan, J.H.L., Chan, D.W.M., Chan, A.P.C. and Lam, P.T.I. (2012) Risk mitigation strategies for guaranteed maximum price and target cost contracts in construction: a factor analysis approach. *Journal of Facilities Management*, 10(1), pp.6–25.

Chan, J.H.L., Chan, D.W.M. and Clifford, B. (2014) New Engineering Contracts (NECs) in practice: empirical evidence from a pilot case study in Hong Kong. *Construction Law Journal*, 30(4), pp.217–235.

Chan, T.K. (2009) Measuring performance of the Malaysian construction industry. *Construction Management and Economics*, 27(12), pp.1,231–1,244.

Chege, L. and Rwelamila, P.D. (2000) Risk management and procurement systems – an imperative approach. In: *Proceedings of the CIB W92 Symposium*, Santiago, Chile, April 2000.

Chen, C.T. and Cheng, H.L. (2009) A comprehensive model for selecting information system project under fuzzy environment. *International Journal of Project Management*, 27(4), pp.389–399.

Cheng, R.L.L. (2004) *Investigation of the Application of Guaranteed Maximum Price in the Hong Kong Construction Industry*. Unpublished BSc (Hons) dissertation in Construction Economics and Management, Department of Building and Real Estate, The Hong Kong Polytechnic University.

Cheung, I. (2008) New Engineering Contract (NEC). *EC Harris Asia Commentary*, July.

Cheung, S.O., Suen, H.C.H. and Cheung, K.K.W. (2004) PPMS: a Web-based construction project performance monitoring system. *Automation in Construction*, 13(3), pp.361–376.

Chevin, D. (1996) The max factor. *Building*, 17 May, pp.40–42.

Chow, L.K. (2005) *Incorporating Fuzzy Membership Functions and Gap Analysis Concept into Performance Evaluation of Engineering Consultants – Hong Kong Study*. Unpublished PhD thesis, Department of Civil Engineering, The University of Hong Kong.

Clough, R.H. and Sears, G.A. (1994) *Construction Contracting*, 6th edition. New York: Wiley-Interscience Publications.

Construction Industry Review Committee (2001) *Construct for Excellence*. Report of the Construction Industry Review Committee, HKSAR.

Cooper, D.F., Grey, S., Raymond, G. and Walker, P. (2005) *Project Risk Management Guidelines: Managing Risk in Large Projects and Complex Procurements*. Chichester: John Wiley & Sons.

Cox, A.W. and Townsend, M. (1998) *Strategic Procurement in Construction: Towards Better Practice in the Management of Construction Supply Chains*. London: Thomas Telford.

Cox, I.D., Morris, J.H.R. and Jared, G.E. (1999) A quantitative study of post contract design changes in construction. *Construction Management and Economics*, 17(4), pp.427–439.

Cox, R.F., Issa, R.J.A. and Ahren, D (2003) Management's perception of key performance indicators for construction. *Journal of Construction Engineering and Management*, 129(2), pp.142–151.

Davies, A. and Mackenzie, I. (2014) Project complexity and systems integration: constructing the London 2012 Olympics and Paralympics Games. *International Journal of Project Management*, 32(2014), pp.773–790.

Davis Langdon and Seah (2003) *Research Study on Alternative Procurement Approaches for Public Construction Works Projects: Final Report.* Research study for the Environment, Transport and Works Bureau of Hong Kong SAR Government, October.

De Marco, A., Briccarello, D. and Rafele, C. (2009) Cost and schedule monitoring of industrial building project: case study. *Journal of Construction Engineering and Management*, 135(9), pp.853–862.

Dunn, M. and Jones, R. (2004) The Tsim Sha Tsui experience – an update. *Seminar on Collaboration not Confrontation in Executing Construction Contracts* organised by the Lighthouse Club, Hong Kong Convention and Exhibition Centre, 21 May 2004.

EC Harris (2009) *Effective Integration and Partnering Delivery T5 on Time, on Budget.* Available at: http://www.britishexpertise.org/bx/upload/Member_pro jects/EC_Harris_T5.pdf. Accessed 27 June 2016.

EC Harris (2011) How to make target cost and cost reimbursable contracts work. *Contract Solutions E-newsletter*, 4.

Edwards, L. (1995) *Practical Risk Management in the Construction Industry.* London: Thomas Telford.

Egan, J. (1998) *Rethinking Construction.* London: Department of Environment, Transport and the Regions.

El-Sayegh, S.M. (2008) Risk assessment and allocation in the UAE construction industry. *International Journal of Project Management*, 26(4), pp.431–438.

Emsley, R., Rabinowitz, J., Torreman, M., Schooler, N., Kapala, L., Davidson, M. and McGory, P. (2003) The factor structure for the Positive and Negative Syndrome Scale (PANSS) in recent-onset psychosis. *Schizophrenia Research*, 61(1), pp.47–57.

Environment, Transport and Works Bureau (2004) *Reference Guide on Selection of Procurement Approach and Project Delivery Techniques, Technical Circular (Works) No. 32/2004.* Hong Kong: HKSAR Government.

Environment, Transport and Works Bureau (2005) *Risk Management for Public Works: Risk Management User Manual.* Hong Kong: HKSAR Government.

Eom, C.S.J. and Paek, J.H. (2009) Risk index model for minimizing environmental disputes in construction. *Journal of Construction Engineering and Management*, 135(1), pp.34–41.

Eriksson, P.E., Atkin, B. and Nilsson, T. (2009) Overcoming barriers to partnering through cooperative procurement procedures. *Engineering, Construction and Architectural Management*, 16(3), pp.598–611.

Fan, A.C.W. and Greenwood, D. (2004) Guaranteed maximum price for the project? *Surveyors Times*, 13(3), pp.20–21.

Ferreira, R.M.L. and Rogerson, J.H. (1999) The quality management role of the owner in different types of construction contract for process plant. *Total Quality Management*, 10(3), pp.401–411.

Flanagan R. and Norman, G. (1993) *Risk Management and Construction*. Oxford: Blackwell Scientific Publications.

Frampton, J. (2003) Can't be too sure on paper. *Sydney Morning Herald*, November.

Gander, A. and Hemsley, A. (1997) Guaranteed maximum price contracts. *CSM*, January, pp.38–39.

Garlick, A.R. (2007) *Estimating Risk: A Management Approach*. Aldershot: Ashgate.

Gorsuch, R.L. (1983) *Factor Analysis*, 2nd edition. Hillsdale, NJ: Erlbaum.

Grove, J.B. (2000) *The Grove Report: Key Terms of 12 Leading Construction Contracts are Compared and Evaluated, Report Prepared by the Author for the Government of the Hong Kong SAR*. Available at: http://www.construc tionweblinks.com/Resources/Industry_Reports__Newsletters/Nov_6_2000/grove_report.htm#endnotes. Accessed 4 September 2006.

Hair, J.F., Anderson, R.E., Tatham, R.L. and Black, W.C. (1998) *Multivariate Data Analysis*, 3rd edition. New York: Macmillan.

Haley, G. and Shaw, G. (2002) Is 'guaranteed maximum price' the way to go? *Hong Kong Engineer*, January.

Haponava, T. and Al-Jibouri, S. (2012) Proposed system for measuring project performance using process-based key performance indicators. *Journal of Management in Engineering*, 28(2), pp.140–149.

Harris, N. (2002) The cost of Wembley has shot up by around £550 million: so who's netting the extra cash? *The Independent*, 26 September. Available at: http://www.independent.co.uk/sport/general/the-cost-of-wembley-has-shot-up-by-acircpound550m-so-whos-netting-the-extra-cash-643745.html. Accessed 4 September 2006.

Hartman, F. (2000) *Don't Park Your Brain Outside: A Practical Guide to Improving Stakeholder Management Value with SMART Management*. Philadelphia, PA: PMI Publications.

Hauck, A.J., Walker, D.H.T., Hampson, K.D. and Peters, R.J. (2004) Project alliancing at National Museum of Australia – collaborative process. *Journal of Construction Engineering and Management*, 130(2), pp.143–152.

Heaphy, I. (2011) Do target cost contracts deliver value for money? Paper presented to the Society of Construction Law, Leeds, 17 May.

HK-BEAM (2005) *An Environmental Assessment Method for Existing Buildings*. Hong Kong Building Environmental Assessment Method (HK-BEAM) Society.

Hong Kong Engineer (2011) All Systems Go for MTR's South Island Line and Kwun Tong Line Extension. *Hong Kong Engineer*, 39(7). Available at http://www.hkengineer.org.hk/program/home/article.php?aid=6065&volid=133. Accessed 16 June 2016.

Hong Kong Housing Authority (2006) *Internal Guidelines for Guaranteed Maximum Price Contract Procurement Based on Private Sector Model*. Hong Kong Housing Authority, HKSAR Government.

Hong Kong Housing Authority (2011) *Planning, Design and Delivery of Quality Public Housing in the New Millennium, Hong Kong*. Hong Kong Housing Authority, HKSAR Government.

Hughes, D., Trefor, W. and Ren, Z. (2012) Is incentivisation significant in ensuring successful partnered projects? *Engineering, Construction and Architectural Management*, 19(3), pp.306–319.

Hughes, W., Kwarwu, W. and Hillig, J.B. (2011) Contracts and incentives in the construction sector. In: Caldwell, N. and Howard, M. (eds) *Procuring Complex Performance*. Abingdon: Taylor & Francis.

Jones, K. and Kaluarachchi, Y. (2008) Performance measurement and benchmarking of a major innovation programme. *Benchmarking: An International Journal*, 15(2), pp.124–136.

Kadefors, A. (2004) Trust in project relationships – inside the black box. *International Journal of Project Management*, 22(3), pp.175–182.

Kaka, A., Wong, C. and Fortune, C. (2008) Culture change through the use of appropriate pricing systems. *Engineering, Construction and Architectural Management*, 15(1), pp.66–77.

Kaluarachchi, Y.D. and Jones, K. (2008) Monitoring of a strategic partnering process: the Amphion experience. *Construction Management and Economics*, 25(10), pp.1,053–1,061.

Kartam, N.A. and Kartam, S.A. (2001) Risk and its management in the Kuwaiti construction industry: a contractors' perspective. *International Journal of Project Management*, 19(6), pp.325–335.

Ke, Y., Wang, S.Q., Chan, A.P.C. and Lam, P.T.I. (2010) Preferred risk allocation in China's public–private partnership (PPP) projects. *International Journal of Project Management*, 28(5), pp.482–492.

Kerzner, H. (1995) *Project Management: A Systems Approach to Planning, Scheduling and Controlling*, 5th edition. New York: Van Nostrand.

Khalfan, M.M.A., McDermott, P. and Swan, W. (2007) Building trust in construction projects. *Supply Chain Management: An International Journal*, 12(6), pp.385–391.

Knight, K. and Fayek, A.R. (2002) Use of fuzzy logic for predicting design cost overruns on building projects. *Journal of Construction Engineering and Management*, 128(6), pp.503–512.

KPI Working Group (2000) *KPI Report for the Minister for Construction*. London: Department of the Environment, Transport and the Regions.

Lam, E.W.M., Chan, A.P.C. and Chan, D.W.M. (2007) Benchmarking the performance of design-build projects: development of project success index. *Benchmarking: An International Journal*, 14(5), pp.624–638.

Lam, K.C., Wang, D., Lee, T.K.P. and Tsang, Y.T. (2007) Modelling risk allocation decision in construction contracts. *International Journal of Project Management*, 25(5), pp.485–493.

Lam, P.T.I. (2002) Buildability assessment: the Singapore approach. *Journal of Building and Construction Management*, 7(1), pp.21–27.

Laryea, S. (2011) Quality of tender documents: case studies from the UK. *Construction Management and Economics*, 29(3), pp.275–286.

Laryea, S. and Hughes, W. (2008) How contractors price risk in bids: theory and practice. *Construction Management and Economics*, 26(9), pp.911–924.

Latham, M. (1994) *Constructing the Team: Final Report of the Joint Government/industry Review of Procurement and Contractual Arrangements in the United Kingdom Construction Industry*. London: HMSO.

Lewis, D. (2002) Dispute resolution in the New Hong Kong International Airport Core Programme Projects – postscript. *International Construction Law Review*, 19(1), pp.68–78.

Lewis, S. (1999) GMP contracts: are they worth the risk? *Construction Law*, 10(3), pp.25–27.

Li, B., Akintoye, A., Edwards, P.J. and Hardcastle, C. (2005) The allocation of risk in PPP/PFI construction projects in the UK. *International Journal of Project Management*, 23(1), pp.25–35.

Lingard, H. and Rowlinson, S. (2006) Letter to the editor. *Construction Management and Economics*, 24(11), pp.1,107–1,109.

Lo, S.M. (1999). A fire safety assessment system for existing buildings. *Fire Technology*, 35(2), pp.131–152.

Lord, W., Liu, A., Tuuli, M.M. and Zhang, S. (2010) A modern contract: developments in the UK and China. In: *Proceedings of the Institution of Civil Engineers – Management, Procurement and Law*, 163(4), pp.151–159.

Lu, R.S., Lo, S.L. and Hu, J.Y. (1999) Analysis of reservoir water quality using fuzzy synthetic evaluation. *Stochastic Environmental Research and Risk Assessment*, 13, pp.327–336.

Luu, V.T., Kim, S.Y. and Huynh, T.A. (2008) Improving project management performance of large contractors using benchmarking approach. *International Journal of Project Management*, 26(7), pp.758–769.

Mahesh, G. (2009) Gain/pain Share and Relational Strategies to Enhance Value in Target Cost and GMP Contracts. Unpublished PhD thesis, Department of Civil Engineering, The University of Hong Kong.

Mass Transit Railway Corporation (2003) *The Tseung Kwan O Extension Success Story*. Hong Kong: Mass Transit Railway Corporation Ltd.

Masterman, J.W.E. (2002) *An Introduction to Building Procurement Systems*, 2nd edition. London: Spon Press.

Menches, C.L. and Hanna, A.S. (2006) Quantitative measurement of successful performance from the project manager's perspective. *Journal of Construction Engineering and Management*, 132(12), pp.1,284–1,293.

Meng, X. (2012) The effect of relationship management on project performance in construction. *International Journal of Project Management*, 30(2), pp.188–198.

Meng, X. and Gallagher, B. (2012) The impact of incentive mechanisms on project performance. *International Journal of Project Management*, 30 (2012), pp.352–362.

Mills, R.S. and Harris, E.C. (1995) Guaranteed maximum price contracts. *Construction Law*, 573(95), pp.28–31.

Ministry of Business Innovation and Employment (2012) *Energy Data File 2012*. Wellington, New Zealand: Ministry of Business Innovation and Employment. Available at http://www.mbie.govt.nz/info-services/sectors-industries/energy/energy-data-modelling/publications/energy-data-file/documents-image-library/energy-data-file-2012.pdf/view. Accessed 16 June 2016.

Minogue, A. (1998) Firms reject GMP contracts. *Building*, May.

Moore, C., Mosley, D. and Slagle, M. (1992) Partnering guidelines for win-win project management. *Project Management Journal*, 22(1), pp.18–21.

Mosey, D (2009) *Early Contractor Involvement in Building Procurement Contracts, Partnering and Project Management*. Chichester: Wiley-Blackwell.

MTRC (2003) *The Tseung Kwan O Extension Success Story*. Hong Kong: Mass Transit Railway Corporation.

Muller, R. and Turner, R. (2010) Leadership competency profiles of successful project managers. *International Journal of Project Management*, 28(5), pp.437–448.

Mylius, A. (2007) Supply management. *Building*, 15 June, pp.20–23.

National Building and Construction Council (1989) *Strategies for the Reduction of Claims and Disputes in the Construction Industry – No Dispute*. Canberra, Australia: National Building and Construction Council.

National Economic Development Office (1982) *Target Cost Contracts: A Worthwhile Alternative*. London: Civil Engineering Economic Development Committee, National Economic Development Office.

Ng, A. and Loosemore, M. (2007) Risk allocation in the private provision of public infrastructure. *International Journal of Project Management*, 25(1), pp.66–76.

Ng, S.T., Xie, J., Skitmore, M. and Cheung, Y.K. (2007) A fuzzy simulation model for evaluating the concession items for public-private partnership schemes. *Automation in Construction*, 17(1), pp.22–29.

NHS (2010) *Executive Summary of ProCure21+*. London: National Health Service, Department of Health. Available at http://www.procure21plus.nhs.uk/executive-summary. Accessed 16 June 2016.

NHS (2011). *The ProCure21+ Guide: Achieving Excellence in NHS Construction*. London: National Health Service, Department of Health. Available at http://www.procure21plus.nhs.uk/resources/downloads/ProCure21Plus%20Guide.pdf. Accessed 16 June 2016.

Nicolini, D., Holti, R. and Smalley, M. (2001) Integrating project activities: the theory and practice of managing the supply chain through clusters. *Construction Management and Economics*, 19, pp.37–47.

Nicolini, D., Tomkins, C., Holti, R. and Oldman, A. (2000) Can target costing and whole life costing be applied in the construction industry? Evidence from two case studies. *British Journal of Management*, 11(4), pp.303–324.

Ojo, A.S. and Ogunsemi, D.R. (2009) Assessment of contractor's understanding of risk management in Seychelles construction industry. In: *Proceeding of RICS COBRA Research Conference*, University of Cape Town, South Africa, 10–11 September 2009.

Olawale, Y.A. and Sun, M. (2010) Cost and time control of construction projects: inhibiting factors and mitigating measures in practice. *Construction Management and Economics*, 28(5), pp.509–526.

Oztas, A. and Okmen O. (2004) Risk analysis in fixed priced design build construction project. *Building and Environment*, 49(2), pp.229–237.

Patterson, L. (1999) Legal – the trouble with GMP. *Building*, 37.

Perkin, T. (2008) *An Evaluation of Guaranteed Maximum Price (GMP) Procurement Strategies and Performance in the South Australian Construction Industry*. BSc dissertation in Construction Management and Economics, School of Natural and Built Environments, University of South Australia.

Perkin, T. and Ma, T. (2010) Guaranteed maximum price (GMP) contracting in the South Australian construction industry. In: *Proceedings of International Conference on Construction & Real Estate Management*, Brisbane, Australia, 1, pp.178–181.

Perry, J.G. and Barnes, M. (2000) Target cost contracts: an analysis of the interplay between fee, target, share and price. *Engineering, Construction and Architectural Management*, 7(2), pp.202–208.

Perry, J.G. and Thompson P.A. (1982) *Target and Cost-reimbursable Construction Contracts*, CIRIA Report R85. London: CIRIA.

Potts, K. and Ankrah, N. (2014) *Construction Cost Management: Learning from Case Studies*. Abingdon: Routledge.

Pryke, S. and Pearson, S. (2006) Project governance: case studies on financial incentives. *Building Research and Information*, 34(6), pp.534–545.

Rahman, M.M. and Kumaraswamy, M.M. (2002) Joint risk management through transactionally efficient relational contracting. *Construction Management & Economics*, 20(1), pp.45–54.

Rojas, E.M. and Kell, I. (2008) Comparative analysis of project delivery systems: cost performance in Pacific Northwest public schools. *Journal of Construction Engineering and Management*, 134(6), pp.387–397.

Rosch, E. (1988) Principle of categorisation. In: Collins, A. and Smith, E. (eds) *Readings in Cognitive Science*. Los Altos, CA: Morgan Kaufman.

Rose, T. and Manley, K. (2010) Client recommendations for financial incentives on construction projects. *Engineering, Construction and Architectural Management*, 17(3), pp.252–267.

Sadiq, R., Husain, T., Veitch, B. and Bose, N. (2005) Risk-based decision-making for drilling waste discharge using fuzzy synthetic evaluation technique. *Ocean Engineering*, 31(16), pp.1,929–1,953.

Sadler, M.C (2004) The Use of Alternative Integrated Procurement Approaches in the Construction Industry. Unpublished MBA dissertation in Construction and Real Estate, Department of Construction Management and Engineering, University of Reading, UK.

Septelka, D. and Goldblatt, S. (2005) *Survey of General Contractor/construction Management Projects in Washington State*. Report to State of Washington Joint Legislative Audit and Review Committee, Seattle, WA.

Shen, L.Y., Platten, A. and Deng, X.P. (2006) Role of public private partnerships to manage risks in public sector projects in Hong Kong. *International Journal of Project Management*, 15(2), pp.101–107.

Shen, Q.P. and Chung, J.K.H. (2006) A critical investigation of the briefing process in Hong Kong's construction industry. *Facilities*, 24(13/14), pp.510–522.

Sidwell, T. and Kennedy, R. (2004) Re-valuing construction through project delivery. In: Khosrowshahi, F. (ed.) *Proceedings of the 20th Annual ARCOM Conference*, 1–3 September 2004, Heriot-Watt University, Edinburgh, UK, 1, pp.55–65.

Singh, B., Dahiya, S., Jain, S., Garg, V.K. and Kushwaha, H.S. (2008) Use of fuzzy synthetic evaluation for assessment of groundwater quality for drinking usage: a case study of Southern Haryana, India. *Environmental Geology*, 54(2), pp.249–255.

Soetanto, R., Proverb, D.G. and Holt, G.D. (2001) Achieving quality construction projects based on harmonious working relationship: clients' and architects' perceptions of contractor performance. *International Journal of Quality and Reliability Management*, 18(5), pp.528–548.

Song, L., Mohamed, Y. and AbouRizk, S.M. (2009) Early contractor involvement in design and its impact on construction schedule performance. *Journal of Management in Engineering*, 25(1), pp.12–20.

Stukhart, G. (1984) Contractual incentives. *Journal of Construction Engineering and Management*, 110(1), pp.34–42.

Swan, W. and Kyng, E. (2004) *An Introduction to Key Performance Indicators*. Salford, UK: Centre of Construction Innovation.

Tah, J.H.M. and Carr, V. (2000) A proposal for construction project risk assessment using fuzzy logic. *Construction Management and Economics*, 18(4), pp.491–500.

Tam, A. (2002) Chater House – central destination. *Building Journal Hong Kong China*, December, pp.28–35.

Tam, V.W.Y., Shen, L.Y., Tam, C.M. and Pang, W.S.P. (2007) Investigating the intentional quality risks in public foundation projects: a Hong Kong study. *Building and Environment*, 42(1), pp.330–343.

Tang, S.L. and Lam, R.W.T. (2003) Applying the target cost contract concept to price adjustments for design-and-build contracts. *Hong Kong Engineer*, September, pp.8–19.

Tang, W.Y. (2005) An Evaluation of the Success and Limitations of Guaranteed Maximum Price in the Hong Kong Construction Industry. Unpublished BSc (Hons) dissertation in Construction Economics and Management, Department of Building and Real Estate, The Hong Kong Polytechnic University.

Taroun, A. (2014). Towards a better modelling and assessment of construction risk: insights from a literature review. *International Journal of Project Management*, 32(1), pp.101–115.

Tay, P., McCauley, G. and Bell, B. (2000) Meeting client's needs with GMP. *The Building Economist*, June, pp.4–5.

Tennant, S. and Langford, D. (2008) The construction project balanced score card. In: Dainty, A. (ed.) *Proceedings of the 24th Annual ARCOM Conference*, 1–3 September 2008, Cardiff, UK, pp.361–370.

Thompson, P. and Perry, J. (1992) *Engineering Construction Risks: A Guide to Project Risk Analysis and Risk Management*. London: Thomas Telford.

Ting, W. (2006) The Impact of the Interdisciplinary Efforts on the Receptivity of Guaranteed Maximum Price (GMP) Project. Unpublished MSc dissertation in Inter-disciplinary Design Management, Department of Real Estate and Construction, The University of Hong Kong.

Toor, S.R. and Ogunlana, S.O. (2010) Beyond the 'iron triangle': stakeholder perception of key performance indicators (KPIs) for large-scale public sector development projects. *International Journal of Project Management*, 28(3), pp.228–236.

Trench, D. (1991) *On Target: A Design and Manage Target Cost Procurement System*. London: Thomas Telford.

Turner, J. and Simister, S. (2001) Project contract management and a theory of organisation. *International Journal of Project Management*, 19(8), pp.457–64.

Uebergang, K., Galbraith, V. and Tam, A.M.L. (2004) *Sustainable Construction – Innovations in Action*. Hong Kong: Civic Exchange.

Walker, D. and Hampson, K. (2003) Procurement choices. In: Walker, D. and Hampson, K. (eds) *Procurement Strategies: A Relationship-based Approach*. Oxford: Blackwell.

Walker, D., Hampson, K. and Peters, R. (2000) *Relationship-based Procurement Strategies for the 21st Century*. Canberra, Australia: AusInfo.

Walker, D., Hampson, K. and Peters, R. (2002), Project alliancing vs project partnering: a case study of the Australian National Museum Project. *Supply Chain Management: An International Journal*, 7(2), pp.83–91.

Ward, S.C., Chapman, C.B. and Curtis, B. (1991) On the allocation of risk in construction projects. *International Journal of Project Management*, 9(4), pp.140–147.

Williams, T., Williams, M. and Ryall, P. (2013) Target cost contracts: adopting innovative incentive mechanisms to improve the project delivery process. *Management*, 759, p.768.

Wong, A.K.D. (2006) The application of a computerised financial control system for the decision support of target cost contracts. *Journal of Information Technology in Construction*, 11, pp.257–268.

Wong, F.W.H., De Saram, D.D., Lam, P.T.I. and Chan, D.W.M. (2006) A compendium of buildability issues from the viewpoints of construction practitioners. *Architectural Science Review*, 49(1), pp.81–90.

Wong, P.S.P. and Cheung, S.O. (2004) Trust in construction partnering: views from parties of the partnering dance. *International Journal of Project Management*, 22(6), pp.437–446.

Wong, P.S.P and Cheung, S.O. (2005) Structural equation model of trust and partnering success. *Journal of Construction Engineering and Management*, 21(2), pp.70–80.

Wright, J.N. and Fergusson, W. (2009) Benefits of the NEC ECC form of contract: A New Zealand case study. *International Journal of Project Management*, 27(3), pp.243–249.

Yeung, J.F.Y., Chan, A.P.C. and Chan, D.W.M. (2009) Developing a performance index for relationship-based construction projects in Australia: Delphi study. *Journal of Construction Engineering and Management*, 25(2), pp.59–68.

Yeung, J.F.Y, Chan, A.P.C., Chan, D.W.M. and Li, L.K. (2007) Development of a Partnering Performance Index (PPI) for construction projects in Hong Kong: a Delphi study. *Construction Management and Economics*, 25(12), pp.1,219–1,237.

Yew, M. (2008) Guaranteed maximum price (GMP) contracts in Singapore. *EC Harris Asia Commentary*, January.

Yiu, C.Y., Ho, H.K., Lo, S.M. and Hu, B.Q. (2005) Performance evaluation for cost estimators by reliability interval method. *Journal of Construction Engineering and Management*, 131(1), pp.108–116.

Yu, A.T.W., Shen, Q.P., Kelly, J. and Hunter, K. (2007) An empirical study of the variables affecting construction project briefing/architectural programming. *International Journal of Project Management*, 25(1), pp.198–212.

Zaghloul, R. and Hartman, F. (2003) Construction contracts: the cost of mistrust. *International Journal of Project Management*, 21(6), pp.419–424.

Zeng, J., An, M. and Smith, N.J. (2007) Application of a fuzzy based decision making methodology to construction project risk assessment. *International Journal of Project Management*, 25(6), pp.589–600.

Zhang, G. and Zou, P.X.W. (2007) Fuzzy analytical hierarchy process risk assessment approach for joint venture construction project in China. *Journal of Construction Engineering and Management*, 133(10), pp.771–779.

Zhang, H., Li, H. and Tam, C.M. (2004) Fuzzy discrete-event simulation for modeling uncertain activity duration. *Engineering, Construction and Architectural Management*, 11(6), pp.426–437.

Zimina, D., Ballard, G. and Pasquire, C. (2012) Target value design: using collaboration and a lean approach to reduce construction cost. *Construction Management and Economics*, 30(5), pp.383–398.

Index